WORKED EXAMPLES IN CHEMICAL REACTION ENGINEERING

By

B. N. Nnolim

ISBN 978-1-906914-19-6

Other Engineering Books by Ben Nnolim Books

Title	**Paperback**	**eBook**
Fundamentals of Mass Transfer	978-1-906914-01-1	978-1-906914-27-1
Worked Examples in Mass Transfer	978-1-906914-46-2	978-1-906914-47-9
Applied Heat Transfer Volume One: Conduction of Heat in Solids (With Worked Examples)	978-1-906914-75-2	978-1-906914-26-4
Applied Heat Transfer Volume 2 (With Worked Examples): Heat Convection in Fluids	978-1-906914-22-6	978-1-906914-25-7
The Development and Analysis of New Chemical Plants and Processes	978-1-906914-48-6	

Ben Nnolim Books,
7 Sandway Path,
St Mary Cray,
Orpington, Kent
BR5 3TS, UK
Email: benedictnnolim@aol.com

DEDICATION

This book is dedicated to my deceased sisters and brother who were true professionals in their own fields

Mrs Regina M. Agazie

Mrs Philomena N. Muoghallu

Chukwueloka Anthony Nnolim Esq

PREFACE

The subject of chemical reactions is so vast and complicated that the mere thought of dealing with it in a book of this type is intimidating. Inorganic and organic chemistry, alone and between them, have thousands of reactions, many with unique and differing characteristics and mechanisms.

The chemical research and development engineer or the industrial process engineer, who has to either design or utilise these reactions, to make products, cannot operate, successfully, without finding a way of grouping and classifying them.

The oldest approach has been to compress this vast amount of knowledge and information on chemical reactions into four main but very simplified areas of (1) chemical kinetics, (2) reaction stoichiometry, (3) reactor design and optimisation and (4) reactor operation and control.

More modern approaches are more holistic and, although all the aspects of the old approach are still used, tend to emphasise the identification of models and simulated entities of the whole reaction system.

This book has followed the old approach essentially because many practitioners in the developing world, to which this book is addressed, are not familiar with the sophisticated mathematics of the new approach of modelling and simulation. Many chemical plants, in their region of the world, usually meant for import substitution manufacturing, are almost always turn-key and, as far as they are concerned, black boxes. The old approach is, also, more useful to them as prospective chemical reactor designers for their indigenous products and processes.

In dealing with problems in chemical kinetics and stoichiometry in this book, an attempt has been made to survey and classify, broadly, the main types and varieties of chemical reactions before explaining the generalised kinetics and stoichiometry which may be applied to them. The approach to reactor design, operation and control, is concerned only with the chemical engineering aspects of the material and energy balances and not with the mechanical, process control or economic aspects which are better treated elsewhere in the specialised literature on the subjects.

Graphical procedures used to be dominant in the analysis of elementary kinetics. They have been de-emphasised here because computer calculations and graphics have more or less replaced manual graph plotting

The original materials in this book come mainly from my lecture notes which, in turn, were compiled from many sources - journal articles, trade literature, textbooks and direct practical experience. The later availability of the internet has enabled me to revise and update some of the information. These sources are acknowledged in the relevant chapters.

I must acknowledge the tenacity and industry of my past students who, in spite of the subject appearing completely alien to their mentality, world view and educational background, struggled enthusiastically and with great faith to comprehend the subject matter.

Finally, I thank God for everything and the Immaculate Conception for her ceaseless intercession.

Benedict Nnolim
March 6, 2012

Table of Contents

CHAPTER ONE: FUNDAMENTAL DEFINITIONS AND CONCEPTS

Example 1.1

What is a chemical reaction?

Answer

A chemical reaction is a process by which the constituent atoms of molecules are rearranged, under suitable conditions, to form new molecules.

In PVT systems, such conditions are defined by the pressure acting on the system (P), system volume (V) and system temperature (T).

Reactants are substances with which a chemical reaction is started, and which are changed on taking part in the reaction.

Products are called those substances which are formed as a result of the reaction and are, in general, different from the reactants.

Example 1.2

What is a chemical equation?

A chemical equation is an identity, in symbolic terms, which describes completely, the chemical reaction. Under the IUPAC (International Union of Pure and Applied Chemists) convention, the reactants are written to the left and the products to the right of the identity or equality sign.

Thus, if 2 units of A react with 3 units of B to form 1 unit of C and 4 units of D, the equation of the chemical reaction would be given as

$$2A + 3B = C + 4D \tag{1.0}$$

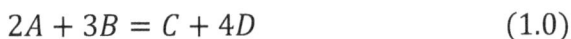

A and B are the *reactants*, C and D, the *products*.

Example 1.3

What is stoichiometry?

Answer

Stoichiometry (sometimes called reaction stoichiometry to distinguish it from composition stoichiometry) is the quantitative relationship between reactants and products in chemical reactions.

This quantitative relationship is illustrated, in the chemical equation, by numbers, called stoichiometric coefficients.

Stoichiometric coefficients, associated with reactants, have a negative sign while those associated with the products have a positive sign.

For example, in the chemical reaction of equation (1.0) above, -2, -3, +1 and +4, are the *stoichiometric coefficients* of the reaction.

The mathematics, used in stoichiometry, is based on the law of conservation of mass, the law of definite proportions or of constant composition and on the law of multiple proportions. The main assumption is that reactants combine in definite ratios of compounds.

Stoichiometry (reaction stoichiometry) is often used to balance chemical equations. For example, the two diatomic gases, hydrogen and oxygen, can combine to form liquid, water, in an exothermic reaction, as described by the following equation:

$$2H_2 + O_2 = 2H_2O \tag{1.1}$$

That is, 4H and 2O on the left hand side must be equal to 4H and 2O on the right hand side, respectively, of equation (1.1).

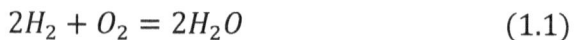

The term stoichiometry is, also, often used to describe the molar proportions of elements in stoichiometric compounds (composition stoichiometry). A stoichiometric compound is one in which the molar proportions of its constituents are whole numbers (the law of multiple proportions). For example, the stoichiometry of hydrogen and oxygen in the stoichiometric compound, H_2O, is 2: 1.

Stoichiometry is used not only to balance chemical equations but also in conversions, such as converting kilograms to kilomoles (kmols), or from kilograms to cubic metres.

Stoichiometry is also used to find the right amount of reactants to use in a chemical reaction

Example 1.4

You have seen the definitions of reaction stoichiometry and composition stoichiometry. What is gas stoichiometry?

Answer

Stoichiometry is referred to as *gas stoichiometry* when it is employed for reactions in which gases are produced. The gases produced are assumed to be ideal, with known temperature, pressure, and volume. Often, but not always, the standard temperature and pressure (STP) are taken as 0°C and 1 atmosphere and used as the conditions for gas stoichiometric calculations. In the petroleum industry, 60 F or 15.5 C and 1 atm are used as standard.

The two common forms of the ideal gas equation used are

$$PV = nRT \tag{1.7}$$

and

$$\frac{P_0V_0}{T_0} = \frac{P_1V_1}{T_1} \tag{1.8}$$

where P, V, T, refer to any pressure, volume or temperature, P_0, V_0, and T_o refer to STP and P_1, V_1 and T_1 refer to a specified condition. R is the gas constant and n, the number of moles of gas present in the volume V.

In gas stoichiometry, it is often desired to know the molar mass, *n*, of a gas of mass, *m*, and known mass density, ρ. In this case, the ideal gas law, $PV = nRT$ is rearranged to obtain a relation between the mass density and the molar mass of the gas, assumed to be ideal. The molar mass, *n*, is given by

$$n = \frac{mass,\ m}{molecular\ weight,\ M} = \frac{m}{\mathrm{M}} \qquad (1.9)$$

and the mass density by

$$\rho = \frac{mass,\ m}{volume,\ V} = \frac{m}{V} \qquad (1.10)$$

Since, from the ideal gas law, equation (1.7), the molar density is given by

$$\rho_M = \frac{n}{V} = \frac{P}{RT} \qquad (1.11)$$

the mass density is related to the molar density, from (1.10), (1.9) and (1.11), as

$$\rho = \frac{m}{V} = \frac{nM}{V} = M\rho_M = \frac{MP}{RT} \qquad (1.12)$$

where (in consistent units):

P = absolute gas pressure
V = gas volume
n = number of moles
R = ideal gas law constant
T = absolute gas temperature
ρ = gas density at T and P
m = mass of gas
M = molar mass of gas

Example 1.5

What other definitions of the stoichiometric coefficient are in use?

Answer

A more general definition of the stoichiometric coefficient, that is useful in both simple and complex chemical reactions, is that the stoichiometric coefficient, ν_i, of the ith component is

$$\nu_i = \frac{dN_i}{d\xi} \quad or \quad dN_i = \nu_i d\xi \tag{1.13}$$

where N_i is the number of molecules of i, and ξ is the progress variable or extent of reaction.

The extent of reaction can be regarded as a real or hypothetical product, one molecule of which is produced each time the reaction event occurs.

When the reaction is analysed in terms of its reaction mechanism, stoichiometric coefficients will always be integers, since elementary reactions always involve whole molecules. When the analysis is based on the overall reaction, some stoichiometric coefficients may turn out to be rational fractions.

Chemical species present during reaction which do not participate in a reaction, such as inerts, have stoichiometric coefficients of zero. Any chemical species which is regenerated, such as a catalyst, also has a stoichiometric coefficient of zero.

In very simple reactions such as the isomerism shown in equation (1.14) below

$$A \leftrightarrow B \tag{1.14}$$

$\nu_B = 1$ since one molecule of B is produced each time the reaction occurs, while $\nu_A = -1$ since one molecule of A is necessarily consumed.

Because the total mass is conserved, as well as the numbers of atoms of each kind, in any chemical reaction, the number of possible values for the stoichiometric coefficients is, similarly, constrained.

In multiple reactions, in which any chemical component can participate in several reactions simultaneously, the stoichiometric coefficient of the ith component in the kth reaction is defined as

$$\nu_{ik} = \frac{dN_i}{d\xi_k} \tag{1.15}$$

The total differential change in the amount of the ith component in all the multiple reactions taking place, thus, becomes

$$dN_i = \sum_k \nu_{ik} d\xi_k \tag{1.16}$$

The extent of reaction, ξ, is, especially, useful in analysing complex reaction systems because it can be used to represent a reaction system both in terms of the amounts of the chemicals present, N_i, as state variables, and in terms of the actual compositional degrees of freedom, expressed as the extent of reaction, ξ_k.

Example 1.6

What is the stoichiometry of the compound whose formula is given as $Na_2CO_3.10H_2O$?

Answer

The stoichiometry is

$$Na: C: O: H :: 2: 1: 4: 20 \qquad Ans$$

Example 1.7

How many moles of NaCl are to be found in 2 kg of NaCl?

Answer

From equation (1.9)

$$n = \frac{m}{\mathrm{M}} = \frac{2}{58.22}\left(\frac{kg}{1}\right).\left(\frac{kmol}{kg}\right) = 0.034\ kmol\ NaCl\ \ \text{Ans}$$

Example 1.8

Estimate the amount of aluminium required to reduce 85 kg of iron (III) oxide to iron according to the reaction shown below

$$Fe_2O_3 + 2Al \rightarrow Al_2O_3 + 2Fe$$

Answer

The formula weights of Fe_2O_3, Al, Al_2O_3 and Fe are, respectively, 159.7, 27, 102, and 55.85.

According to the stoichiometry of the given equation, 1 kmol (or 159.7 kg) of Fe_2O_3 reacts with 2 kmol (or 54 kg) of Al, to produce 1 kmol (or 102 kg) of Al_2O_3 and 2 kmol (or 111.7 kg) of Fe . It follows, therefore, that 85 kg Fe_2O_3 will react with

$$\frac{54}{159.7} x\ 85 = 28{,}74\ kg\ Al \qquad\qquad Ans$$

Example 1.9

The overall reaction for the Solvay process is

$$CaCO_3 + 2NaCl = Na_2CO_3 + CaCl_2$$

How many kilograms of limestone ($CaCO_3$) and salt ($NaCl$) are required to produce 5000 kg of soda ash (Na_2CO_3)?

Answer

Stoichiometry requires that 1 kmol of $CaCO_3$ will react with 2 kmols of $NaCl$ to produce 1 kmol of Na_2CO_3.

The formula weight of $CaCO_3$ is 40+12+48 = 100; that of $NaCl$ 23+35.5 = 58.5 and that of Na_2CO_3 is 46+12+48 = 106.

We can use stoichiometry to assert that if

$$100\ kg\ of\ CaCO_3\ reacts\ with\ 2\ x\ 58.5\ kg\ NaCl\ to\ form\ 106\ kg\ Na_2CO_3$$

Then, the amount of $CaCO_3$ required to produce 5000 kg Na_2CO_3 will be

$$\frac{5000\ x\ 100}{106} = 4716.98\ kg \quad Ans$$

Similarly, the amount of $NaCl$ required to produce 5000 kg Na_2CO_3 will be

$$\frac{5000\ x\ 2\ x\ 58.5}{106} = 5518.87\ kg \quad Ans$$

Example 1.10

For the reaction

$$CH_4 + H_2O \leftrightarrow CO + 3H_2$$

The rate of production of H_2 in a reactor operated at steady state is 6 kmol/h. What are the rates of consumption/production of CH_4, H_2O and CO?

Answer

The stoichiometry of the reaction indicates that 1 kmol of CH_4 will react with 1 kmol of H_2O to produce 1 kmol CO and 3 kmol of H_2. This is the so called steam reforming reaction for the production of hydrogen.

If the production rate of H_2 is 6 kmol/h, then the consumption of CH_4 and H_2O will, each be 2 kmol/h while the production rate of CO will be 2 kmol/h. Ans.

Example 1.11

Calculate the volume, at STP, of gaseous NO_2 produced from the combustion of 100 kg of NH_3, by the reaction:

$$4NH_3\ (g) + 7O_2\ (g) \rightarrow 4NO_2\ (g) + 6H_2O(l)$$

What would this volume be at at a temperature of 50 C and 5 atmospheres pressure?

Answer

The number of kmol of of $NH_3(g)$ equivalent to 100 kg of $NH_3(g)$ is given by

$$n = \frac{m}{M} = \frac{100}{17}\left(\frac{kg}{1}\right) \text{ x } \left(\frac{1\ kmol}{kg}\right) = 5.882\ kmol\ NH_3\ (g) \quad (1)$$

Since the combustion of 1 kmol of NH_3 (g) produces 1 kmol of NO_2 (g) , 5.882 kmol $NH_3\ (g)$ would, also, produce 5. 882 kmol of NO_2.

Since 1 kmol of an ideal gas, at STP, occupies 22.42 m^3, 5.882 kmol of NO_2 will occupy 5.882 x 22.42 = 131.87 m^3.

At 50 C = 273 + 50 = 323 K and 5 atm pressure, from equation (1.8)

$$\frac{P_0V_0}{T_0} = \frac{P_1V_1}{T_1}$$

Thus

$$\frac{1 \text{ x } 22.42}{273} = \frac{5 \text{ x } V_1}{323}$$

from which

$$V_1 = 5.305\ m^3\ \ Ans$$

Example 1.12

Balance the following chemical reaction

$$HNO_3 + CuS = Cu(NO_3)_2 + S + H_2O + NO$$

using the most general method that you know.

Answer

The balanced equation will have stoichiometric coefficients such That

$$\nu_1 HNO_3 + \nu_2 CuS = \nu_3 Cu(NO_3)_2 + \nu_4 S + \nu_5 H_2O + \nu_6 NO \qquad (1)$$

such that

$$\sum_j \nu_j C_j = 0 \qquad (2)$$

and

$$\sum_{j=1}^{N} \nu_j n_j^k = 0 \qquad (3)$$

where

C_j	is the jth component, reactant or product
ν_j	is the stoichiometric coefficient of component, reactant or product j
n_j^k	is the number of atoms of the kth element in the jth component, reactant or product

Thus, Table 1.12, below, can be built up from equations (1).

From Table 1.12 and equation (3), the following equations can be derived

$$\nu_1 - 2\nu_5 = 0 \qquad (4)$$

$$\nu_1 - 2\nu_3 - \nu_6 = 0 \qquad (5)$$

$$3v_1 - 6v_3 - v_5 - v_6 = 0 \tag{6}$$

$$v_2 - v_3 = 0 \tag{7}$$

$$v_2 - v_4 = 0 \tag{8}$$

Table 1.12: Stoichiometric Coefficients for Equation (1)

	HNO_3	CuS	$Cu(NO_3)_2$	S	H_2O	NO
Hydrogen atoms	v_1				$-2v_5$	
Nitrogen atoms	v_1		$-2v_3$			$-v_6$
Oxygen atoms	$3v_1$		$-6v_3$		$-v_5$	$-v_6$
Copper atoms		v_2	$-v_3$			
Sulphur atoms		v_2		$-v_4$		

From equations (7) and (8),

$$v_2 = v_3 = v_4 \tag{9}$$

From equation (4)

$$v_5 = \frac{v_1}{2} \tag{10}$$

From equations (6) and (10)

$$3v_1 - 6v_3 - \frac{v_1}{2} - v_6 = 0$$

That is

$$5v_1 - 12v_3 - 2v_6 = 0 \tag{11}$$

From equation (5) multiplied by 2

$$2v_1 - 4v_3 - 2v_6 = 0 \tag{12}$$

Subtracting equation (11) from equation (12)

$$8v_3 = 3v_1$$

from which

$$v_3 = \frac{3}{8}v_1 \qquad (13)$$

Substituting equation (13) in equation (5)

$$v_1 - 2\,x\,\frac{3}{8}v_1 - v_6 = 0$$

from which

$$v_6 = \frac{1}{4}v_1 \qquad (14)$$

Thus, from equations (9), (10), (13), (14) and (1)

$$v_1 HNO_3 + \frac{3}{8}v_1 CuS = \frac{3}{8}v_1 Cu(NO_3)_2 + \frac{3}{8}v_1 S + \frac{v_1}{2}H_2O + \frac{1}{4}v_1 NO \qquad (15)$$

If we multiply equation (15) across by $\frac{8}{v_1}$, we get the balanced equation which, also, satisfies equation (2)

$$8HNO_3 + 3CuS = 3Cu(NO_3)_2 + 3S + 4H_2O + 2NO \qquad Ans$$

References for Chapter One

1. Hougen O. A., Watson K. M and Ragatz R. A.*; Chemical Process Principles Part 1: Material and Energy Balances*; 2nd Edition; John Wiley and Sons Inc., N. Y. USA, (1954)

2. Myers A. I, Seider W. D.; *Chemical Engineering and Computer Calculations*; Prentice Hall, N. J. USA (1976).

3. Schmidt A. X, List H. L.; *Material and Energy Balances*; Prentice Hall, N. J. , USA, (1962

CHAPTER TWO: TYPES OF CHEMICAL REACTIONS

Example 2.1

Outline, briefly, the major types of chemical reactions in current use.

Answer

Chemical reactions may be described in different ways depending on the mechanism of the reaction, the phase of reaction, the energy requirements for the reaction, the nature of the reacting materials and products and on the branch of chemistry or biochemistry (inorganic, organic, polymer, electrochemical, biochemical, etc) which dominates the reaction. None of these excludes the others

The most useful descriptions are based on proven utility and industry wide acceptance of the description. Some examples of widely used terms for describing common kinds of reactions are:

Isomerisation, in which a chemical compound undergoes a structural rearrangement without any change in its net atomic composition;

Direct combination or synthesis, in which two or more chemical elements or compounds unite to form a more complex product:

$$N_2 + 3H_2 \rightarrow 2NH_3 \quad (2.1)$$

Chemical decomposition or analysis, in which a compound is decomposed into smaller compounds or elements:

$$2H_2O \rightarrow 2H_2 + O_2 \quad (2.2)$$

Single displacement or substitution, characterized by an element being displaced out of a compound by a more reactive element:

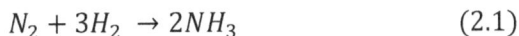

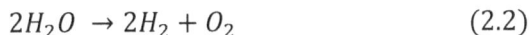

$$2Na + 2HCl \rightarrow 2NaCl + H_2 \tag{2.3}$$

Double displacement or coupling substitution, in which two compounds in aqueous solution (usually ionic) exchange elements or ions to form different compounds:

$$NaCl + AgNO_3 \rightarrow NaNO_3 + AgCl \tag{2.4}$$

Combustion, in which any combustible substance combines with an oxidizing element, usually oxygen, to generate heat and form oxidized products.

$$C_{10}H_8 + 12O_2 \rightarrow 10CO_2 + 4H_2O \tag{2.5}$$

$$CH_2S + 6F_2 \rightarrow CF_4 + 2HF + SF_6 \tag{2.6}$$

The term combustion is, normally, used on large-scale oxidation of whole molecules. Thus, the controlled oxidation of a single functional group is not combustion.

Some kinds of reactions have similarities which make it possible to define some larger groups. A few examples are:

Inorganic reactions involve mainly inorganic compounds

Organic reactions involve compounds which have carbon as the main element in their molecular structure. These reactions occur mostly among, according to, within, by, or via, functional groups.

Petrochemical reactions are often distinguished from other organic reactions even though most petrochemical compounds are, indeed, organic compounds.

Redox reactions involve augmenting or decreasing the electrons associated with a particular atom according to its oxidation number.

Catalytic reactions in which the rate of a chemical reaction is influenced by a substance (catalyst) which may or may not change chemically during the course of reaction.

Electrochemical reactions which are electrically or chemically driven and may be used either to synthesise, or decompose, chemical compounds or to produce electric power.

Polymerisation reactions whose unique characteristic is the high molar mass involved when compared to normal chemical reactions. *Enzymatic (Fermentation) reactions* which are reactions catalysed by enzymes, and, up to the state of present knowledge, take place in biological systems.

When the phase in which chemical reactions take place becomes influential the reaction may be said to be *homogenous* or *heterogeneous* without excluding any other classification.

Similarly, reactions may be considered *exothermic* or *endothermic* depending on whether heat energy is released or absorbed during reaction.

Other general classifications include *reversible reactions* (rates of the forward reaction and backward reaction are significant or comparable.) and *irreversible reactions* (only the forward reactions are significant or dominant).

There are, also, descriptions of chemical reactions as zero, first, second or third order reactions and as chain, nuclear, radical and clock reactions.

Example 2.2

What is an elementary reaction? How does it differ from a complex reaction?

Answer

An *elementary reaction* is the simplest possible reaction between the smallest units (radicals, ions, atoms, molecules, etc) of reactants through the shortest possible step or path. It takes place in one step as indicated by the stoichiometric equation.

The IUPAC Compendium of Chemical Terminology defines an elementary reaction as a reaction for which no reaction intermediates have been detected or need to be postulated in order to describe the chemical reaction on a molecular scale.

An elementary reaction is assumed to occur in a single step and to pass through a single transition state.

Elementary reactions, as opposed to *stepwise reactions*, have these distinctive features: *stoichiometry* (the numbers of particles in the reaction equation), *molecularity* (the actual number of molecules colliding) and *reaction order* which must coincide with molecularity.

The rate equation for an elementary reaction, is easily written, by merely looking at the reaction equation, as the product of a constant (the reaction rate constant) and the concentration of the reacting species raised to the power of the corresponding stoichiometric coefficients. That is

$$Rate, \qquad r = kC^{\nu} \tag{2.7}$$

where k is a constant, C a concentration and v a power to which the concentration is raised

The reaction equation for elementary reactions coincides with the process taking place at the atomic level, i.e. *a* molecules of type A are colliding with *b* molecules of type B (*a* plus *b* is the molecularity).

When the chemical reaction consists, however, of a sequence or combination of elementary reactions, it is a *non-elementary* or *complex reaction.*

Example 2.3

Describe, briefly, the most commonly encountered types of complex chemical reactions

.Answer

The most frequently encountered complex reactions are

- Consecutive reactions (also called series reactions)
- Simultaneous reactions (also called parallel, sometimes, competing reactions)
- Opposing reactions (also called equilibrium reactions)
- Multiple reactions which are often combinations of two or more elementary and non-elementary reactions.
- Chain reactions
- Nuclear chain reactions
- Clock reactions

Consecutive (Series) Reactions

Reactions are said to be consecutive if they follow one another. Consecutive reactions are important in auto-catalysed and similar reactions.

$$A \rightarrow B \rightarrow C \tag{2.8}$$

Competing Reactions

Reactions are competing or parallel if any one or more of the reactants or products takes part in another, simultaneously, occurring reaction. For example, a parallel reaction may take place as follows:-

$$\begin{gathered} A + B = C \\ C + D = E \end{gathered} \tag{2.9}$$

or as

$$\begin{gathered} A + B = C \\ C + D = E \\ A + C = E \end{gathered} \tag{2.10}$$

(reactants B and C are in competition to react with A)

or as

$$\begin{gathered} A \xrightarrow{k_1} B \\ A \xrightarrow{k_2} C \end{gathered} \tag{2.11}$$

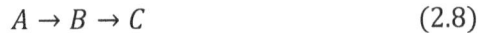

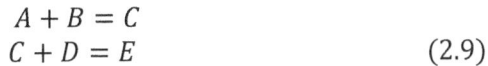

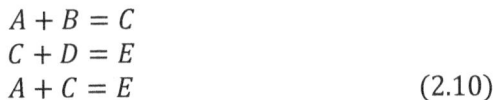

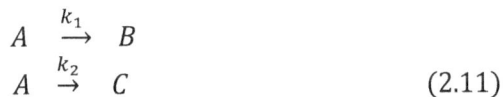

Parallel reactions are important in commercial processes where they, often, pose problems of selecting the preferred product or reaction path (selectivity).

Opposing (Equilibrium) Reactions

Reactions are said to be opposed to each other when some of the reactions take place in the forward direction while others proceed in the backward direction. For example

$$A + B \overset{k_1}{\rightarrow} C + D$$
$$C + D \overset{k_2}{\rightarrow} A + B \qquad (2.12)$$

Chain Reactions

Chain reactions are those reactions whose features are most easily explained by a chain mechanism. A typical chain mechanism may consist of:-

- an initiation step
- a propagation step or steps
- inhibition
- termination

The reaction between hydrogen and bromine to form hydrogen bromide (HBr) is explained by a chain mechanism:

Initiation $\quad Br_2 \overset{k_1}{\rightarrow} Br^* + Br^*$

Propagation $\quad H_2 + Br^* \overset{\mathrm{kp}_2}{\longrightarrow} HBr + H^*$

Inhibition $\quad H^* + HBr \overset{\mathrm{kp}_4}{\longrightarrow} H_2 + Br^*$

Termination $\quad Br^* + Br^* \overset{\mathrm{k}_{-1}}{\longrightarrow} Br_2 \qquad (2.13)$

It is not necessary that every chain reaction displays all the four characteristics listed. Chain reactions are more common with photochemical and free radical reactions and often help to explain the mechanisms of very fast reactions and explosions.

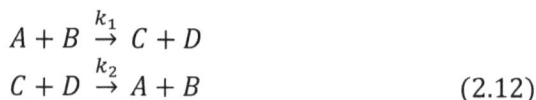

Nuclear Chain Reaction

A nuclear *chain reaction* occurs when, on average, more than one nuclear reaction is caused by another nuclear reaction, thus leading to an exponential increase in the number of nuclear reactions.

Clock Reactions

A *chemical clock* is a complex mixture of reacting chemical compounds in which the concentration of one component shows an abrupt change accompanied by a visible colour effect. The onset of the colour change may be used to tell time.

Clock reactions are useful in the study of complex reactions or natural metabolic and biochemical processes. In a clock reaction, the main reaction can be monitored by another reaction involving one or more of its reactants or products and further still indicated by another reaction. The purpose is to obtain a rate expression for the main reaction.

Examples of clock reactions are the *Belousov-Zhabotinsky* reaction, the *Bray-Liebhafsky* reaction and the *iodine clock* reaction. These are *oscillatory reactions* in which the concentration of products and reactants can be approximated in terms of damped oscillators.
Most clock reactions involve three steps

The Main Reaction

For example $A + B = C + D$

such as the oxidation of persulphate ions by iodide

$$2I_{(aq)} + S_2O^{2-}_{8(aq)} = 2SO^{2-}_{4(aq)} + I_{2(aq)} \qquad (2.14)$$

The Monitor Reaction (which must be faster than the main reaction)

For example $C + M \rightarrow Products$

where $C = I_2\ (aq)$ and $M = S_2O_3^{2-}$ (thiosulphate ion), the monitor substance. The initial concentration of the monitor substance must be less than those of the main reactants A and B

The Indicator Reaction (which must also be faster than the main reaction)

For example: $C + Indicator\ 1 = Indicator\ 2$

where *Indicator 1* and *Indicator 2* represent the two indicator colours involved.

Example 2.4

Describe, briefly, the phenomenon of equilibrium reaction and outline its major features.

Answer

Equilibrium reactions tend to occur whenever there is simultaneous generation of products and reactants during a chemical reaction.

In such cases, forward reactions (consuming reactants to form products) and backward reactions (consuming products to form reactants) are significant or comparable.

A major feature of such reactions is the state of chemical equilibrium. In this state, the forward and reverse reaction rates are equal, thus preserving the amount of reactants and products.

A reaction in equilibrium can, however, be driven in the forward or reverse direction by changing the reaction conditions such as temperature or pressure. Le Chatelier's principle can be used to predict whether products or reactants will be formed.

A state of chemical reaction equilibrium is often described by a parameter, called the equilibrium constant, which is unique, for such reactions, at any given temperature and pressure.

The equilibrium constant is, usually, represented by K, and is the ratio of the product of fugacity of the products, normalised to the standard state, to that of the reactants. The equilibrium constant may, also, be expressed in terms of partial pressures, mole fractions or in terms of concentration.

Example 2.5

Derive the commonly encountered expressions for the equilibrium constant of chemical reactions

Answer

The equilibrium constant of chemical reactions, K, may be derived as follows:

Consider any reaction such as

$$aA + bB \rightarrow cC + dD \qquad (2.15)$$

The change in the Gibb's free energy, ΔG_i for each reactant or product, i, at the temperature T, is given by, for gas phase reactions,

$$\Delta G_i = \Delta G_i^0 + R\,T \ln f_i^{\nu_i} \qquad (2.16)$$

where f_i is the fugacity of component i, ΔG_i^0 is the change in the Gibb's free energy at the chosen standard state and ν_i is the stoichiometric coefficient of component, i. If the reaction takes place in the liquid or solid phase, activity, a_i, replaces fugacity in equation (2.16).

Applying these to the reaction (equation 2.15)

$$\Delta G_{reactants} = \Delta G_{reactants}^0 + R\,T \ln f_A^a\, f_B^b \qquad (2.17)$$

$$\Delta G_{products} = \Delta G^0_{products} + R\,T \ln f_C^c f_D^d \qquad (2.18)$$

At equilibrium

$$\Delta G_{reactants} = \Delta G_{products} \qquad (2.19)$$

That is

$$\Delta G^0_{Reaction} = \Delta G^0_{products} - \Delta G^0_{reactants} \qquad (2.20)$$
$$= R\,T \ln f_A^a f_B^b - R\,T \ln f_C^c f_D^d$$

$$= -RTln\,\frac{f_C^c f_D^d}{f_A^a f_B^b} = -RTln\,K_f \qquad (2.21)$$

where K_f is the equilibrium constant at the temperature, T, defined in terms of fugacity. It can be seen that

$$K_f = \frac{f_C^c f_D^d}{f_A^a f_B^b} \qquad (2.22)$$

A more rigorous expression of gas fugacity normalises each fugacity with respect to the standard state of the component to which it applies and results in the following expressions:

$$\Delta G^0_{Reaction} = -RTln\,\frac{\left(\frac{f}{f^0}\right)_C^c \cdot \left(\frac{f}{f^0}\right)_D^d}{\left(\frac{f}{f^0}\right)_A^a \left(\frac{f}{f^0}\right)_B^b} = -RTln\,K_f \qquad (2.23)$$

where

$$K_f = \frac{\left(\frac{f}{f^0}\right)_C^c \cdot \left(\frac{f}{f^0}\right)_D^d}{\left(\frac{f}{f^0}\right)_A^a \cdot \left(\frac{f}{f^0}\right)_B^b} \qquad (2.24)$$

If total system pressure is less than 20 atmospheres, the fugacity can be replaced, without great loss of accuracy, by the partial pressures of the reactants and products to give an equilibrium constant, based on partial pressures, K_P, as

$$K_p = \frac{p_C^c \cdot p_D^d}{p_A^a \cdot p_B^b} \qquad (2.25)$$

and

$$K_p = \frac{\left(\frac{p}{p^0}\right)_C^c \cdot \left(\frac{p}{p^0}\right)_D^d}{\left(\frac{p}{p^0}\right)_A^a \cdot \left(\frac{p}{p^0}\right)_B^b} \tag{2.26}$$

p^0 is the partial pressure at the standard state, equal to one atmosphere.. For reactions in liquids and solids, the replacement of fugacity by activity results in the following expressions for the equilibrium constant.

<u>Equilibrium Constant, K_C, based on the Concentration</u>

Since activity, a = γC, where γ is the activity coefficient and C is the molar concentration, when γ = 1, as happens in dilute solutions,

$$K_C = \frac{C_C^c \, . \, C_D^d}{C_A^a . \, C_B^b} \tag{2.27}$$

When γ ≠ 1, as happens in concentrated solutions, activity, *a*, is used. That is

$$K_a = \frac{a_C^c a_D^d}{a_A^a a_B^b} = \frac{(\gamma C)_C^c . (\gamma C)_D^d}{(\gamma C)_A^a . (\gamma C)_B^b} \tag{2.28}$$

Equation (2.28) is, usually, simplified to

$$K_a = \frac{\gamma_C^c \gamma_D^d}{\gamma_A^a \gamma_B^b} \cdot \frac{C_C^c C_D^d}{C_A^a C_B^b} = K_\gamma \, . \, K_C \tag{2.29}$$

where K_γ is an equilibrium constant based on activity coefficients.

<u>Equilibrium Constant, K_Y, based on Mole Fractions</u>

In the gas and liquid phases, the mole fractions, *y* and *x* are defined as

$$y = \frac{p}{P_T} \quad (gas\ phase) \tag{2.30}$$

$$x = \frac{C}{C_T} \quad (liquid\ phase) \tag{2.31}$$

where P_T and C_T are the total system pressure and concentration, respectively.

The equilibrium constant, based on mole fractions, K_Y or K_X would, then, be defined as

$$K_y = \frac{y_C^c \cdot y_D^d}{y_A^a \cdot y_B^b} \tag{2.32}$$

$$K_x = \frac{x_C^c . x_D^d}{x_A^a . x_B^b} \tag{2.33}$$

For ideal gases,

$$f_i = p_i = y_i P_T = C_i RT \tag{2.34}$$

$$K_P = K_f = K_C RT^{\delta} = K_y P_T^{\delta} \tag{2.35}$$

where

$$\delta = \frac{d}{a} + \frac{c}{a} - \frac{b}{a} - 1 \tag{2.36}$$

For solid components,

$$\frac{f}{f^0} = 1 \tag{2.37}$$

Example 2.6

What is the effect of temperature on the equilibrium constant

Answer

The equilibrium constant is a function of temperature only. This is illustrated by the equation, known as the van't Hoff's equation, (equation 2.38 below)

$$\frac{dlnK_P}{dT} = \frac{\Delta H_R}{RT^2} \tag{2.38}$$

The Kirchoff's equation is given by

$$\Delta H_{R_2} = \Delta H_{R_1} + \int_{T_1}^{T_2} \Delta Cp dT \qquad (2.39)$$

When $\Delta Cp = 0$, $\Delta H_{R_2} = \Delta H_{R_1}$ and from equation (2.38)

$$ln\frac{K_{P_2}}{K_{P_1}} = \frac{\Delta H_R}{R}\left(\frac{1}{T_1} - \frac{1}{T_2}\right) \qquad (2.40)$$

When $\Delta Cp \neq 0$, also from equation (2.38)

$$ln\frac{K_{P_2}}{K_{P_1}} = \frac{1}{R}\int_{T_1}^{T_2} \frac{\Delta H_R}{T^2} dT \qquad (2.41)$$

Example 2.7

Explain the term catalysis and illustrate the manner in which catalysts affect chemical reactions.

Answer

Catalysis is the process by which the rate of a chemical reaction is influenced by a substance (catalyst) which may or may not change chemically during the course of reaction. Usually, the catalyst forms an intermediate with the reactant which, in turn, interacts with other reactants, or itself, to form products and regenerate the catalyst. The net result is that the reaction proceeds faster than an uncatalysed reaction, by a mechanism involving lower activation energy.

Catalysis can be classified as:

- *homogenous*, if the catalyst, reactants and products are in the same physical state such as the gaseous or liquid state.
- *heterogeneous*, if the catalyst phase is, distinctively, different from that of the reactants and products. The majority of heterogeneous catalysts are solid (Harshaw Chem. Co., 1980).

Although the manner in which catalysts affect chemical reactions is still, largely, an art and empirically based, in specific cases, , certain general explanations have, however, been accepted with respect to the manner in which catalysts accelerate the rate of chemical reactions and determine selectivity of certain reactions over others.

This state of affairs may be summarised in Figures 2.1 and 2.2 which show the potential energy of the reactants at their initial state, the potential energy of the transition state or activated complex and the potential energy of the products or the final state for the same reaction when it is uncatalysed and when it is catalysed.

Fig. 2.2 shows that the formation of an intermediate compound at *B* is associated with less activation energy than that of the uncatalysed reaction. Note that, for the same reaction, the heat of reaction is still the same for both the catalysed and uncatalysed reaction.

Fig. 2.1: Potential Energy Map of Uncatalysed Reaction

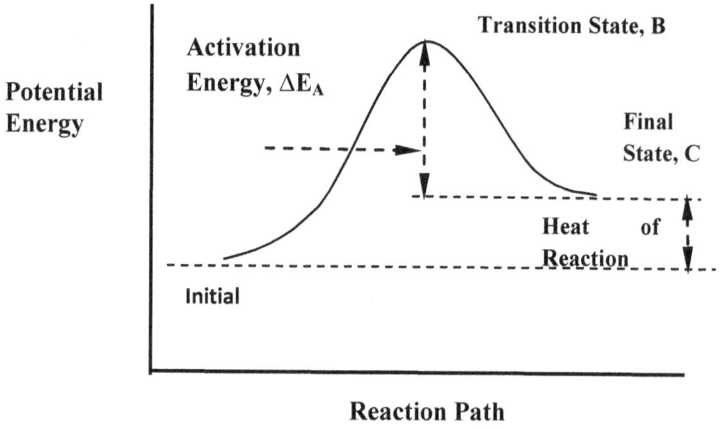

A catalyst promotes the preferential selection of a particular reaction by lowering the activation energy of that reaction as shown in Fig. 2.2 more than that of others. Since the rate constant k, is given by equation (2.24) as

$$k = f(T) = k_0 e^{-\frac{\Delta E_A}{RT}} \qquad (2.42)$$

it is easy to see that a reduction of ΔE_A will result in a higher k which is equivalent to a faster reaction.

The drop in energy of activation is governed by the formation of chemical bonds between the reaction intermediates and the catalyst. It is found that at low bond strengths, formation of these bonds between the reaction intermediates and the catalyst is the rate determining step whereas at high bond strengths, it is the dissociation of the complex, formed by the chemical bonding of the reaction intermediates to the catalyst, that is, the rate determining step. This implies that high bond strength is not necessarily desirable at all times. The experience is that good selectivity is obtained, for any set of catalysts and reactions, when the bond strength is within a finite range.

Fig. 2.2: Potential Energy Map of the Catalysed Reaction

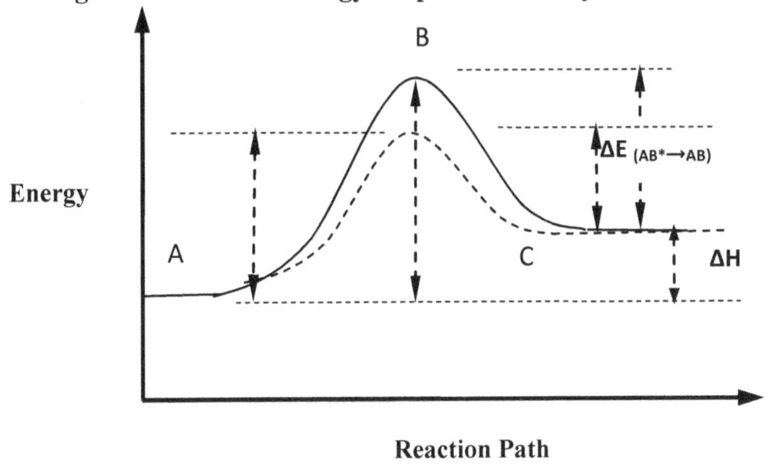

Example 2.8

Distinguish between homogeneous and heterogeneous catalysis giving relevant examples of each.

Answer

Homogenous catalysis occurs when the catalyst and the reactants are of the same phase, usually, liquid or gas.

Gas phase catalysed reactions are, usually, first order. A typical example is the oxidation of SO_2 to SO_3 over an N_2O catalyst. That is:

$$\frac{1}{2}O_2 + SO_2 \xleftrightarrow{N_2O\ catalyst} SO_3 \qquad (2.43)$$

with a rate equation

$$-r_{SO_2} = -\frac{d[SO_2]}{dT} = k[SO_2] \qquad (2.44)$$

Other examples include the gas phase decomposition of acetaldehyde, formaldehyde, methanol, ethylene oxide, aliphatic esters, with molecular iodine as catalyst.

Liquid phase catalysed reactions are of two types:

- *Acid – Base catalysed reactions* such as the inversion of sugars, hydrolysis of esters and amides, halogenations of acetone and nitro-paraffins, mutarotation of glucose, esterification of alcohols and the enolisation of aldehydes and ketones. The rate constant is given by the Bronsted equation,

$$k = CK^a \qquad (2.45)$$

where k = the catalytic rate constant
K = the ionisation constant of acid or base
C = a constant dependent only on temperature, solvent or reaction type
a = another constant whose value is between 0.3 and 0.9

- *Metal ion catalysed reactions* in which metal ion, in solution, forms a complex of the chelate type such as *FeEDTA*.

Many more examples of homogeneous catalysts are listed in Table 2.8a.

Homogeneous catalysts, generally, suffer from the disadvantage of considerable problems with catalyst recovery and separation from the products of the reaction.

Heterogeneous catalysis form the more important industrial application of catalysis. This is because the separation of a solid catalyst from gaseous reaction products is, generally, easier than separation of solid catalysts from liquids (eg. the filtration after centrifugation of solid catalyst from hardened fats).

Some examples of heterogeneous catalysis are to be found in the manufacture of organic and mineral acids, ammonia, methanol, and, also, in many cracking, alkylation, polymerisation reactions. Other examples are given in Table 2.8b.

Table 2.8a: Homogeneous Catalysts by Reaction Type (Spex Industries, 1982)

Reaction Type	Catalysts
Acetylenic substitution	$Pd(O_2CCH_3)_2$
Alkylation	$Pd^0(dba)_2$
Aromatisation	$IrCl(CO)(Ph_3P)_2$
Arylation	$Pd(O_2CCH_3)_2; PdCl_2$
Asymetric Hydroformylation	$PtCl_2; RhH(CO)(Ph_3P)_3$
Auto-oxidation	$PdCl_2$
Carbonylation	$Pd(O_2CCH_3)_2; Pd(C_5H_7O_2)_2; PdCl_2;$ $PdCl_2(Ph_3P)_2; H_2PtCl_6.6H_2O; PtCl_2$
Carboxylation	$Pd^0(dba)_2$
Cyclization	$Pd(C_5H_7O_2)_2; PdCl_2$
Co-trimerization	$Pd(O_2CCH_3)_2$
Decarbonylation	$RhCl(Ph_2PCH_2CH_2P\,Ph_2)_2; RhCl(Ph_3P)_3$
Dimerization	$Pd((CC_6H_5)_3P)_4; Pd(O_2CCH_3)_2;$ $Pd(C_5H_7O_2)_2; PdCl_2; PdCl_2(Ph_3P)_2$
Hydration	$PdCl_2$

Hydrogenation	$PdCl_2$; $Ru(CO)_3(Ph_3P)_2$; $RuCl_2(Ph_3P)_3$; $RhH(Ph_3P)_4$; $RhCl(Ph_3P)_3$; $RhH(CO)(Ph_3P)_3$; $RuH_2(PPh_3)_4$; $RuCl_2(Ph_3P)_3$; $RuCl_3.3H_2O$; $Cr(CO)_6$; $CH_3OOCC_6H_5Cr(CO)_3$
Hydroformylation	$Pt(C_5H_7O_2)_2$; $PtCl_2$; $PtCl_2(Ph_3P)_2$; $RhH(CO)(Ph_3P)_3$; $RhCl(CO)(Ph_3P)_2$; $RhCl(1,5-C_8H_{12})_2$; Rh
Hydrosilylation	$RhCl(Ph_3P)_3$
Isomerization	$PdCl_2(C_6H_5CN)_2$; $RhCl(Ph_3P)_3$; $RuHCl(CO)(Ph_3P)_3$; $Ru(C_5H_7O_2)_3$; $IrCl(CO)(Ph_3P)_2$
Nitration	$Pd(O_2CCH_3)_2$
Oxidation	$RuCl_3.3H_2O$
Reduction	$Pd(O_2CCH_3)_2$; $RuCl_2(Ph_3P)_3$; $RuHCl(CO)(Ph_3P)_3$
Telomerization	$Pd((C_6H_5)_3P)_4$; $Pd(O_2CCH_3)_2$; $Pd(C_5H_7O_2)_2$; $PdCl_2(Ph_3P)_2$; $Pd^0(dba)_2$
Trimerization	$Pd(O_2CCH_3)_2$; Rh

Table 2.8b: Heterogeneous Catalysts by Reaction Type (Spex Industries, 1982)

Reaction Type	Catalyst
Amination	Pd/Alumina
Cleavage	Pd. Pd/Carbon
Dehydrogenation	Pd/Carbon
Dimerization	Rh, Rh/Carbon
Homologation	$RuO_2.xH_2O$; RuO_2
Hydroformylation	Rh
Hydrogenation	PtO_2; Pd/Carbon; Pd/Alumina; Pd/Barium Sulphate; Pd/Calcium Carbonate; Pd^0Hydrate; Pd^0Anhydrous; PdO_2; Pt/Carbon; PtO (Adam's Catalyst); Rh/Alumina; Rh/Carbon; RuO_2; Ru/Alumina; Ru/Carbon
Hydrogenolysis	Pd/Carbon; Pd/Barium Sulphate; Pd/Barium Carbonate; Pd/Calcium Carbonate; Pd Black;

	Pt/Alumina; PtO_2; PtO (Adam's Catalyst); Rh/Alumina
Isomerization	Pd/Carbon
Oxidation	PtO (Adam's Catalyst)
Photocatalytic dissociation of Water	$RuO_2.xH_2O$; RuO_2
Reduction	Pd; Pd /Carbon
Reductive Alkylation of Amines	PtO (Adam's Catalyst)
Syngas Homologation	RuO_2
Trimerization	Rh; Rh/Support
Water Gas shift Reaction	Pt/Alumina

The nature of these catalysts is proprietary but some are described in patents and in trade literature. Their mode of action, however, can be summarised as follows:

- Reactants move from bulk of the fluid phase to the catalyst surface
- Reactants are adsorbed on the catalyst surface
- Adsorbed reactants combine to form products
- Products are desorbed
- Products diffuse away from the surface into the bulk of the fluid

From these it can be seen that the factors which will affect the rate of such catalytic reactions are:

- Mass transfer factors
- Heat transfer factors
- Diffusion properties of reactants and products
- Catalyst properties such as particle size, surface area
- Activation energy for adsorption and desorption of reactants and products, respectively
- Activation energy for the reaction of adsorbed reactants

The manner in which the activation energy of adsorbed reactants affects the rate of catalysed reactions has, already, been discussed.

It is, also, to be noted that the energy for adsorption and desorption of reacting intermediates on the catalyst surface would be expected to affect the rate of reaction. It is a fact that the bonding energy to the catalyst surface and the heat of adsorption, generally, decrease with surface coverage as shown in Fig. 2.8a.

This is because, unlike in homogeneous catalysis, the heat of formation of the surface complex with the reaction intermediates is not uniform but varies across the catalyst surface. If it is assumed that the activation energy for an adsorbed complex is proportional to the adsorption energy, it can be seen, from Fig. 2.8b, that surface *b* which requires less adsorption energy (hence less activation energy) than surface *a* for the same coverage, will require a much smaller coverage to reduce the activation energy of the catalysed reaction. Put another way, surface *b* can adsorb a larger number of active species than surface *a* thus resulting in a higher pre-exponential factor. Both of these mean a larger reaction rate constant.

Fig. 2.8a: Surface Coverage vs Heat of Adsorption (Harshaw Catalysts, 1980)

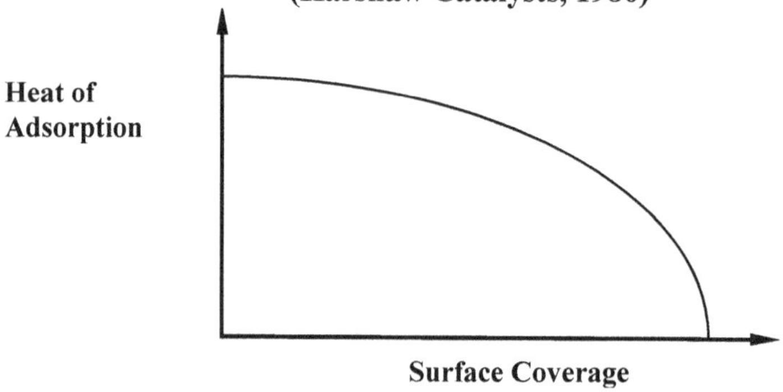

There are catalytic reactions which are not very sensitive to the structure of the active sites. They are known as facile or structure-insensitive reactions. Noble metal catalysts tend to be facile in chemical reactions.

Fig. 2.8b: Effect of Heat of Adsorption on Surface Coverage (Harshaw Catalysts, 1980)

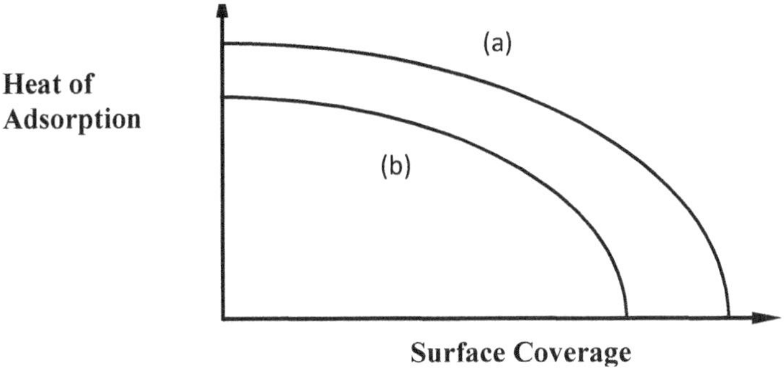

Example 2.9

Outline, with examples, the major characteristics and industrial properties of catalysts of importance in the chemical industry.

Answer

The major characteristics of catalysts may be summarised as follows

1. A catalyst is unchanged at the end of a reaction even though it may undergo changes in the intermediate steps of the reaction.

2. When more than one mechanism is involved, a catalyst may exhibit selectivity over a particular mechanism.

3. The rate of reaction is, usually, proportional to catalyst concentration or surface area.

4. In a reversible reaction, the catalyst affects both the forward and backward reactions equally so that the equilibrium composition is unchanged.

5. A small amount of catalyst must be present in an autocatalytic reaction for the reaction to proceed.

High selectivity is important in chemical reactions where some reaction products are less valuable, more detrimental to personnel or to the environment, or more expensive to separate than others.

Increased rate of reaction is important, first of all, to minimise the size, and hence the cost, of the reactor or reaction process. It is very important in those cases where the uncatalysed reaction proceeds too slowly at temperatures where the equilibrium is most favourable or where it is desirable to carry out the reaction. Here, the catalyst accelerates the rate of reaction while maintaining the favourable equilibrium.

There are cases where, at elevated temperatures, the rate of reaction is favourable but the thermodynamic equilibrium has shifted to a position which restricts conversion. A catalyst is useful in this situation for maintaining acceptable reaction rates at those temperatures which support more favourable equilibrium.

An example is the oxidation of SO_2 to SO_3 by oxygen where a platinum catalyst is used, at about 300 C, to solve the problem of good rates and an unfavourable equilibrium or favourable equilibrium and poor reaction rates.

Because catalysts have to be used in practice, certain properties have to be considered if those catalysts are to be of any commercial use. These properties help to answer questions such as:

- What weight of catalyst will be required to load a specific volume of reactor space?
- How much reactor loading or upsets can the catalyst particle withstand?
- What is the diffusion rate and factor dependence of reactants and products into and out of catalyst pores?
- What is the weight of the active metal charged to a specific volume?

The properties that answer these questions are summarised

according to the treatment given in the trade literature, *Harshaw Catalysts* by the Harshaw Chemical Company (1980)

Apparent Bulk Density (ABD)

This is the measured weight per unit bulk volume of catalyst. Packed ABD is calculated from the weight obtained by filling a graduated cylinder in small increments, tapping the cylinder on a soft surface after each addition until a constant volume results. Loose ABD is calculated from the weight obtained by gently filling a graduated cylinder from a funnel without shaking or tapping.

In commercial reactors, the recommended filling weight for large diameter reactors (small surface area to volume) is 90% of the packed ABD while for small diameter, multi-tube reactors (large surface area to volume ratio) the recommended filling weight is the mean of the loose and packed ABD.

Erosion or Abrasion

This is the weight percentage of fines generated when a stainless steel tube is partially filled with a weighed sample of catalyst, mounted on a revolving drum and rotated for a given time. It is an indication of the attrition expected of the catalyst in commercial use.

Crushing Strength

This is the force required to collapse a single tablet of catalyst. It is an indication of the ability of the catalyst to retain its geometrical shape under load of its neighbours in a commercial reactor.

Micro Mesh Sieve Analysis

This is applied to powders finer than 100 μm in diameter. It is common practice to use a dry sieving method although full wet screening is applied when necessary.

Porosity

This is the ratio of the internal void space within a solid particle to the total external volume of the solid excluding the external dead space among particles.

The internal void space is equal to the difference between the wet drained (in water) and the dry weight of the sample divided by the density of water, assuming its specific gravity to be 1.0.

The total external volume of the solid is the difference between the total volume of the catalyst and water, in which it is completely immersed and boiled to remove air, and the free water poured out after boiling.

The porosity obtained in this manner is the volume porosity and is approximate. It does not include porosities less than 10^{-3} μm diameter.

The weight porosity is given by:

$$\% \, \boldsymbol{WP} = \frac{\boldsymbol{W - D}}{\boldsymbol{D}} \, \boldsymbol{x\, 100} \qquad (2.46)$$

where WP = Weight porosity
W = Wet drained weight of sample
D = Dry weight of sample.

Pore Volume

This is determined, practically, as the volume of water absorbed per unit weight of sample, or more accurately, by

$$\boldsymbol{Pore\,Volume} = \boldsymbol{Mercury\,Volume} - \frac{\boldsymbol{1}}{\boldsymbol{Helium\,density}} \qquad (2.47)$$

where mercury volume, in cm^3/g, is the total external volume of the catalyst determined in liquid mercury. The reciprocal of the helium density is the total volume of the solid part of the catalyst.

Pore Volume Distribution

This is determined by measuring the relative amounts of nitrogen adsorbed or desorbed at different absolute pressures.

Surface Area

The surface area of a catalyst is determined according to the single point method of Brunauer, Emmett and Teller (Single Point BET). The BET equation, for the pressure and the amount of gas adsorbed, is given as:

$$\frac{P}{V(P_0 - P)} = \frac{1}{V_m C} + \frac{C - 1}{V_m C} \cdot \frac{P}{P_0} \qquad (2.48)$$

where P = adsorption pressure
P_0 = saturation pressure of the gas at the temperature of the sample
V = volume of gas (at 0 C, 1 atm.) adsorbed per gram of sample at pressure, P.
V_m = volume of gas (at 0 C, 1 atm.) required for a monolayer
C = constant

In the BET method, a plot of $P/\{V(P_0 - P)\}$ against P/P_0 yields a straight line when P/P_0 is less than 0.35.

In the single point BET method, the intercept, $1/V_mC$, is assumed to be zero. Thus, knowing C, P_0 and measuring P and V, V_m is determined.
Using the accepted value for the area covered by one molecule of nitrogen (0.162 nm^2), the surface area of the catalyst can be determined.

For nickel catalysts, chemisorption of hydrogen at 100 μm Hg pressure is used. At this pressure, monolayer coverage is assumed to be complete. Each hydrogen atom is adsorbed by one nickel atom and occupies 6.5 x 10^{-8} μm^2 area.

Helium Density

A helium-air pycnometer is used to determine the density, in g/cm^3, of the skeleton of the porous material. A pre-dried sample is placed in the sample chamber which is first evacuated then filled with helium. The chamber pressure is decreased to a fixed value in such a way that, depending on the make of the instrument, the volume occupied by the sample can be measured.

The procedure is repeated with a steel sphere of known volume and from this can be computed the absolute volume of the sample material. Helium gas is used because it is an ideal gas which is capable of penetrating into very narrow pores without adsorption on to the surface of the pores. A typical helium-air pycnometer is that made by Micrometrics Instrument Corporation in USA.

Mercury Density – Mercury Pore Volume

A very high pressure porosimeter is used in determining the mercury density and the mercury pore volume of the catalyst.

A suitable sample of the catalyst is weighed and placed in a penetrometer (a glass tube with graduated capillary stem). The penetrometer is placed in a filling device and placed under vacuum so that it is filled with mercury. The mercury density is calculated from the weight of the sample divided by the volume of mercury displaced by the sample.

The penetrometer is then transferred to a pressure chamber where very high pressure can be applied to the mercury. The smallest pore diameter entered by the mercury, under pressure, is determined from

$$D = \frac{175}{P} \qquad (2.49)$$

where D = diameter of the pore in microns (micrometers)
P = absolute pressure in pounds per square inch.

In the instance reported by Harshaw Chem. Co (1980), an Aminco 15000 *psig* porosimeter was capable of measuring pore sizes down to 0.012 μm.

Particle Size

A sedigraph, or particle size analyser by sedimentation, measures the particle size of powder catalysts, in microns, by determining the sedimentation rates of particles in suspension and presenting the data as cumulative mass per cent distribution, assuming Stoke's law and an equivalent spherical diameter.

In a particular instrument made by Micrometrics Instrument Corporation, the concentrations of particles remaining at decreasing sedimentation depths are determined as a function of time using a collimated beam of X-rays. The logarithm of the difference in transmitted X-ray intensity is, electronically, generated, scaled and presented linearly as cumulative mass per cent on the y-axis of an X-Y recorder. The size range handled is, typically, between 50 and 0.18 microns.

Accelerated Attrition Test (For Fluid Bed Catalysts)

This is meant for catalysts which can be suspended in a flow of air and subject to attrition in the process. The test is based on the change in the fraction of particles passing through a 20 micron (micrometers) sieve.

800 cc of humidified catalyst is attrited for one hour in a 5ft x 1 inch ID column using a flow rate of 7.7 litres per minute. A screen test is run on the original sample and on samples after attrition using 20, 45, 60 and 90 micron sieves.

$$\boldsymbol{Attrition\ rate = \frac{100\,A}{B}} \qquad (2.50)$$

where A = increase in % of 0 – 20 micron in the charge
B = % of 20+ microns, in the charge

Thermal Analysis

Differential Scanning Calorimetry (DSC) is used to measure temperature and heat changes associated with transitions in materials. This gives qualitative and quantitative information on physical and chemical changes in the material during endothermic and exothermic processes.

Differential Thermal Analysis (DTA) provides the temperatures at which heat related transitions occur but not the same quantitative heat flow measurements as DSC.

Thermogravimetric Analysis (TGA) measures the weight or weight change of a material continuously, either as a function of temperature or as a function of time, in a desired environment.

Elemental Analysis

Actual chemical analysis of catalysts is done using one of the following methods:

- Classical wet chemical procedures
- Atomic absorption spectrometry (absorption or emission)
- X-ray spectroscopy (fluorescence)
- Emission spectroscopy (quantitative and qualitative)

Other methods for specific or detailed analysis of catalysts or catalytic reactions and products are

1. ESCA (Electron Spectroscopy for Chemical Analysis)
2. Electron Microscopy/SEM(Scanning Electron Microscope)
3. X-ray Diffraction
4. IR (Infra-red) Spectroscopy
5. NMR – ESR (Nuclear Magnetic Resonance – Electron Spin Resonance)

6. Light Microscopy
7. MS, GC, GC-MS and HPLC (Mass Spectrometry, Gas Chromatography, Gas Chromatography – Mass Spectrometry, High Pressure Liquid Chromatography)

Example 2.10

Outline, briefly, the main characteristics and types of polymerisation reactions

Answer

The first unique characteristic of polymerisation reactions is the high molar mass involved when compared to normal chemical reactions. Most polymers of industrial interest have high molar masses between 10^4 and 10^7 and consist of mixtures with a definite distribution of molar mass (Gerrens, 1982).

The high molar mass and precipitous changes in molar mass (up to 10^6 fold), during polymerisation reactions, give rise to very high viscosities and viscosity changes during reaction (Nishimura, 1966). This introduces problems that are much more acute than those observed in low molar mass systems.

Also, polymerisation, as a process, gives rise to a decrease in the entropy of the system. Since

$$\Delta G = \Delta H - T\Delta S \tag{2.51}$$

and ΔG has to be negative for the polymerisation to proceed, the right hand side of equation (2.51) must, therefore, be negative. That is

$$\Delta H = -\Delta G + T\Delta S \tag{2.52}$$

This means that polymerisation processes are, in general, associated with negative ΔH, that is, they are exothermic. This fact, together with the high viscosities and associated large viscosity changes, make heat transfer, reaction control and product quality difficult to manage.

Another unique feature of polymerisation reactions is that, because they are chain reactions, the concentration of chain carriers, ($\approx 10^{-8}$ mol/litre) can be of the same order of magnitude as that of impurities which either terminate the reactions prematurely or hinder them.

Partially or completely synthesized polymer products cannot be reacted further or reprocessed, in the sense of re-polymerisation or reaction, as can be done with low molar mass compounds.

Polymerisation reactions may, also, be classified according to their phase of reaction, as either homogeneous or heterogeneous but it is their kinetics which emphasises their uniqueness in comparison to ordinary chemical reactions.

Polymer reaction kinetics lead to the classification of polymerisation reactions, listed in Fig. 2.10 and Tables 2.10, showing chain reaction mechanisms (addition polymerisation) and step growth (condensation polymerisation) mechanisms.

Three major classes emerge namely monomer linkage without termination, monomer linkage with termination and polymer linkage.

Table 2.10: Classification of Polymerisation Reactions (Gerrens, 1982)

	Phase Based Classification		**Kinetics based Classification**
	Heterogeneous	**Homogeneous**	
1	-	Solution Polymerisation	Monomer linkage, with termination
2	Precipitation Polymerisation	-	Monomer linkage with termination
3	Bead (Suspension)	-	Monomer linkage with termination
4	Emulsion (Suspension)	-	Monomer linkage with termination
5	-	Solution polymerisation	Monomer linkage without termination

6	Precipitation (Suspension)	-	Monomer linkage without termination
7	-	Solution or Melt Polycondensation	Polymer linkage
8	Interfacial polycondensation	-	Polymer linkage
9	Solid phase polycondensation	-	Polymer linkage

Fig. 2.10: Polymerisation Processes (Iqbal, M. Z, 2009)

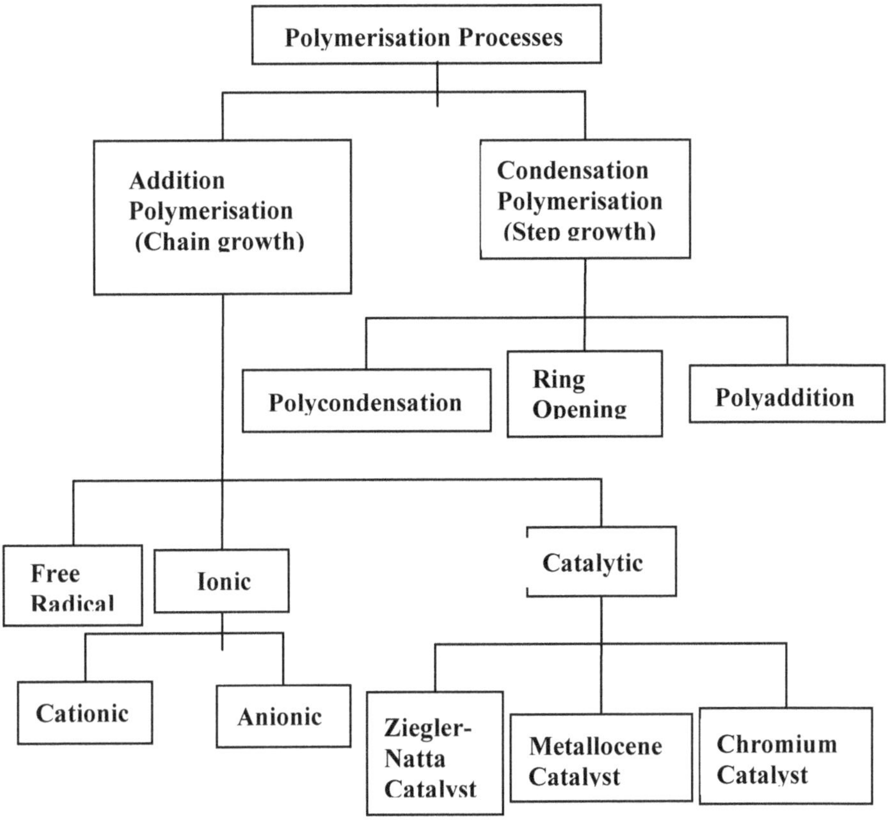

Ref:http://polymer.w99of.com/images/polymerisation processes.jpg

Example 2.11

Outline, briefly, the main characteristics and types of enzymatic reactions

Answer

Enzymatic reactions are reactions catalysed by enzymes, and, up to the state of present knowledge, take place in biological systems. The substance which undergoes transformation (equivalent to the reactants in chemical systems) is referred to as the substrate. The enzyme is, usually, a high molecular weight, natural, protein-like substance having specific catalytic properties.

Enzymes belong to three main groups (Novo Industrie, 1986). The most prevalent group is the *hydrolytic enzymes* or those which hydrolyse or break-up large biopolymers into smaller units. Typical examples of hydrolytic enzymes are proteases, which break up proteins into polypeptides and amino acids. Others are amylases, pectinases, cellulases and lipases which, respectively, break down amyloses, pectins, cellulose and lipoproteins.

The other group of enzymes is the *non-hydrolytic* group. Enzymes in this group include those such as glucose oxidase (which scavenge oxygen in fruit juice, oxidises glucose in egg whites before spray drying), glucose isomerase (which rearranges glucose to form fructose), and aspartate ammonia lyase (used in industrial production of aspartic acid).

The third group of enzymes is the *co-factor* or *co-enzymes* group which, though not enzymes as such, must be present in order to activate the enzyme (Fogler, 1974). Co-factors are substrates which attach to enzymes and are reduced or oxidised during the course of a reaction. They are required for the break down or synthesis of many bio-chemicals such as fatty acids and amino acids.

There are also three broad classes of enzyme reactions. When phase considerations are important (Fogler, 1974) these are:-

1. homogeneous enzyme reactions, in which a soluble enzyme catalyses transformations in a soluble substrate.
2. heterogeneous enzyme reactions in which a soluble enzyme catalyses transformations in an insoluble substrate.
3. heterogeneous enzyme reactions in which an insoluble enzyme catalyses transformations in a soluble substrate.

Type 1 reactions, above, are the most pervasive in nature since almost every synthetic or degradation reaction in living cells, whether of humans, animals or plants, occurs by enzymatic action

In industry, however, type 3 reactions are preferred, followed by type 2 since type 1 reactions, in commercial use, would require expensive separations or enzymes that do not interfere with product quality or performance when discharged with the product.

Heterogeneous enzymatic reactions are very important industrially and are implemented using immobilised enzymes.

When enzymatic reactions are considered in kinetic terms, a classification which is based on the general concept of respiration is used. Respiration is an energy producing process in which organic or reduced inorganic compounds are oxidised by inorganic compounds (Bailey and Ollis, 1977). When the inorganic oxidant is oxygen, the respiration will be *aerobic* while if the oxidant is not oxygen, the respiration is said to be *anaerobic*. When respiration involves an enzyme and occurs under anaerobic conditions, it is called fermentation. Carbohydrates form the most important nutrients during fermentations followed by amino acids.

Fermentations are of more commercial interest and can again be classified in terms of phase, kinetics etc. Glycolysis or homolactic fermentation is a class of fermentations in which a glucose substrate is broken down into lactic acid. This is important in the milk products, cheese, and yoghurt industry.

In alcoholic class of fermentations, alcohols, instead of acids, are produced. This class is of industrial importance in connection with the manufacture of ethyl alcohol from glucose.

Example 2.12

Outline, briefly, the main characteristics and types of electrochemical reactions used in the chemical industry

Answer

Electrochemical reactions are reactions which are electrically or chemically driven and may be used either to synthesise, or decompose, chemical compounds or to produce electric power. Like ordinary chemical reactions, electrochemical reactions may be described as homogenous, heterogeneous, catalytic, non-catalytic, consecutive, parallel, etc.

These classifications, however, are not very useful in the analysis of electrochemical reactions. Parameters of specific relevance to electrochemistry are used. The process of using electric power to synthesise or decompose chemical compounds is known as electrolysis. When chemical reactions are used to produce electric power the result is an electrochemical cell such as a battery or a fuel cell.

Electrochemical reactions can be used in three different ways either as, an electrolytic cell, as a fuel cell or as an electro generator.

Electrochemical Reactions as an Electrolytic Cell

In this case, electric power is used to drive the chemical reaction in order to produce useful chemicals. Typical industrial and successful applications of electrochemical reactions in this manner are in the manufacture, by electrolysis, of chlorine, chlorate, perchlorate, caustic soda, aluminium, hydrogen and oxygen from brine solutions (Dotson, 1978). The cells used are the mercury cell (Castner – Kellnar cell, 1892), the diaphragm cell (Greisheim cell, 1885), and the membrane cell (1970).

The basic reactions, common to all the three types of electrochemical cells, are that, in solution, sodium chloride dissociates into sodium and chloride ions. At the anode, chloride ions are oxidised and chlorine (Cl_2) is formed.

$$2Cl^- \rightarrow Cl_2 + 2e \qquad (2.53)$$

At the cathode, water molecules are reduced to form hydroxyl anions and hydrogen gas

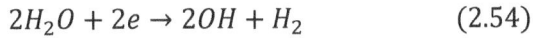

$$2H_2O + 2e \rightarrow 2OH + H_2 \qquad (2.54)$$

The sodium ions in solution and the hydroxyl ions produced at the cathode combine to form sodium hydroxide.

The difference between the three types of cells lies in the manner in which they prevent the sodium and hydroxyl ions from mixing until they can be combined to yield pure sodium hydroxide as

$$Na^+ + OH^- \rightarrow NaOH \qquad (2.55)$$

In the mercury cell, flowing mercury forms an amalgam of Na (Hg) with the sodium ions at the cathode

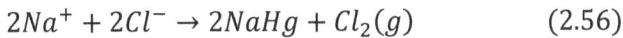

$$2Na^+ + 2Cl^- \rightarrow 2NaHg + Cl_2(g) \qquad (2.56)$$

which is carried to a decomposer where the amalgam reacts with water to release the mercury and form sodium hydroxide

$$2NaHg + 2H_2O \rightarrow 2Na^+ + 2OH^- + H_2(g) + 2Hg \qquad (2.57)$$

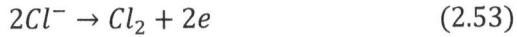

In both the diaphragm and membrane cells, the separation of anode and cathode fluids (anolyte and catholyte, respectively) is achieved by the use of a diaphragm or membrane. These prevent, to varying extents, the migration of chloride ions from the anode to the cathode or hydroxyl ions from cathode to anode, allowing only sodium anions to go from anode to cathode.

The anode, in all the cells, was previously made of graphite, but, now, is made of titanium coated with an electrocatalytic layer of mixed oxides, usually ruthenium oxide (RuO_2) and titanium oxide (TiO_2) (IPPC, 2001). The current trend is to use the so called dimensionally stable anodes (DSA), usually of proprietary composition (Bommaraju et al, 2007).

The cathode is, typically, steel in diaphragm cells, nickel in membrane cells and mercury in mercury cells, often coated with catalytic materials that increase surface area, reduce over voltage or are more stable than the substrate. Such coatings include Ni-S, Ni-Al, Ni-NiO mixtures (IPPC, 2001)

The diaphragm, in the diaphragm cell, is asbestos, vacuum deposited on the cathode and, prevents the migration of hydroxyl ions from the cathode to the anode while allowing the flow and electromigration of sodium ions from the anode to the cathode compartment.

The membrane, in the membrane cell, is a water impermeable, bi-layer ion exchange membrane of perfluorocarboxylic acid and perfluorosulfonic acid films sandwiched between the anode and cathode. It does a better job of preventing the migration of chlorine into the cathode than the diaphragm cell producing less sodium chloride contaminated sodium hydroxide.

Because mercury and asbestos are established health and environmental hazards, the move has been away from mercury and asbestos diaphragm cells (which currently operate with special exemptions by US environmental authorities), to non-asbestos diaphragm and membrane cells (Bommaraju et al, 2007)

The overall reactions in these cells are summarised below while the comparison of the various cell technologies is, also, summarised in Table 2.12a.

For the Mercury Cell

In the electrolyser:

$$Na^+ + Cl^- \leftrightarrow \frac{1}{2}Cl_2 + Na(Hg) \qquad (2.58)$$

Na (Hg) is a sodium – mercury amalgam.

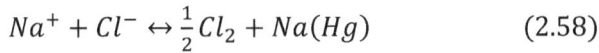

In the decomposer:

$$Na(Hg) + H_2O \leftrightarrow NaOH + \frac{1}{2}H_2 \qquad (2.59)$$

For the Diaphragm and Membrane cells, the electrode reaction is

$$Na^+ + Cl^- + H_2O \leftrightarrow NaOH + \frac{1}{2}H_2 + \frac{1}{2}Cl_2 \qquad (2.60)$$

In the chlorate and hypochlorite industry, it is desirable that the chlorine and caustic, produced by electrolysis, should mix to produce the chlorate. Electrochemical cells, for the manufacture of chlorate, have no mercury and no diaphragm but incorporate cooling coils. This is because the chlorate reactions are very exothermic and slow.

For hypochlorite (bleach) production, saturated brine solution is not required as even a weak solution of common salt will do. The important point is to maintain the pH of the salt solution between 10 and 12 (Bommaraju et al, 2007).

The reactions involved are (Bommaraju et al, 2007)

Electrochemical

At the anode, chloride ions are oxidised and chlorine (Cl_2) is formed.

$$2Cl^- \rightarrow Cl_2 + 2e \qquad (2.52)$$

At the cathode, water molecules are reduced to form hydroxyl anions and hydrogen gas

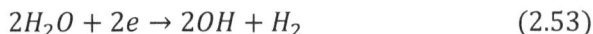

$$2H_2O + 2e \rightarrow 2OH + H_2 \qquad (2.53)$$

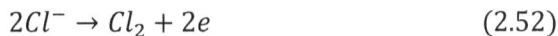

Hypochlorite formation

$$Cl_2 + 2OH^- \leftrightarrow OCl^- + Cl^- + H_2O \tag{2.54}$$

Chlorate formation

$$3OCl^- \leftrightarrow ClO_3{}^- + 2Cl^- \tag{2.55}$$

Overall hypochlorite reaction

$$NaCl + H_2O \rightarrow NaOCl + H_2 \tag{2.56}$$

Overall chlorate reaction

$$NaCl + 3H_2O \rightarrow NaClO_3 + 3H_2 \tag{2.57}$$

<u>Chemical Chlorate formation</u>

$$3Cl_2 + 6NaOH \rightarrow NaClO_3 + 5NaCl + 3H_2O \tag{2.58}$$

Chlorate formation is promoted by the use of saturated brine, acidic solution and temperature close to the boiling point of the solution. Hypochlorite formation is promoted by the use of weak brine, basic solution and low cell temperatures.

Electrochemical Reactions as a Fuel Cell

The fuel cell is a development and an advance on the regular battery in both of which chemical reactions are exploited for the production of electric power. A fuel cell is, however, a battery that has a continuous supply of reactants.
Thus, while battery power is limited to the amount of chemicals in a fixed quantity of a battery pack, the fuel cell generates electric power continuously as long as the reactant chemicals are supplied continuously to the cell, thus maintaining constant chemical potential driving force for the reaction.

This is advantageous because it enables constant power supply to be obtained without the intermediate combustion step and the associated loss of efficiencies, necessary with fired fossil fuel electric power generation systems. In general, fuel cells convert inexpensive chemicals (H_2, CO, O_2) to low value products (H_2O, CO_2).

Fuel cells are classified by the type of electrolyte they use or by the range of temperatures they operate in. Earlier classifications distinguished two main classes of fuel cells, namely the liquid electrolyte (or low temperature) and the solid electrolyte (or high temperature) fuel cell. Modern developments still retain the low temperature/ high temperature classifications but include, in the low temperature classification, solid polymer electrolyte fuel cell which, with greater knowledge, has become known as proton exchange membrane fuel cell.

The high temperature classification includes Solid Oxide Fuel Cell (SOFC) where the electrolyte is a yttria stabilised zirconia ceramic and the Molten Carbonate Fuel Cell (MCFC) in which the electrolyte is a lithium potassium carbonate salt (Wikipedia, 2011).

Typical electrochemical reactions are

Low Temperature Fuel Cells (< 200 C)

At the anode

$$2H_2 \rightarrow 4H^+ + 4e^- \tag{2.59}$$

At the cathode

$$O_2 + 4H^+ \rightarrow 2H_2O \tag{2.60}$$

Overall cell reaction

$$2H_2 + O_2 \rightarrow 2H_2O \tag{2.61}$$

High Temperature Fuel Cells (Wikipedia, 2011)

Solid Oxide Fuel Cell (SOFC) (800 – 100 C)

At the anode

$$2H_2 + 2O^{2-} \rightarrow 2H_2O + 4e^- \tag{2.62}$$

At the cathode

$$O_2 + 4e^- \rightarrow 2O^{2-} \tag{2.63}$$

Overall cell reaction

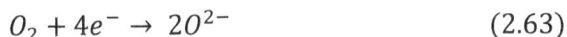

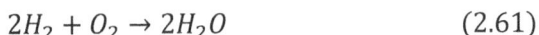

$$2H_2 + O_2 \rightarrow 2H_2O \quad (2.64$$

Molten Carbonate Fuel Cell (MCFC), 650 C

At the anode

$$H_2 + CO_3{}^{2-} \rightarrow H_2O + CO_2 + 2e^- \quad (2.65)$$

At the cathode

$$CO_2 + \frac{1}{2}O_2 + 2e^- \rightarrow CO_3{}^{2-} \quad (2.66)$$

Overall cell reaction

$$H_2 + \frac{1}{2}O_2 \rightarrow H_2O \quad (2.67)$$

A typical fuel cell produces about 0.6 to 0.7 volts per cell with voltage decreasing as current draw increases. Table 2.12a summarises the general characteristics of the more common fuel cell systems while Table 2.12b compares their performance characteristics.

Electrochemical Reactions in Co-generation/ Electrogeneration

By using the fuel cell as a chemical generator, both electrical energy and desirable valuable chemical products can be produced. If a suitable reaction can be utilised in the fuel cell, electricity can be obtained as a by-product from the negative free energy change of the reaction

Electrogenerative processes generally incorporate two coupled electrode reactions, an appropriate barrier electrolyte, provision for product recovery, and a circuit with load for using the electrical energy or possibly only dissipating it (Langer et al, 1986)

Table 2.12a: Comparison of Fuel Cell Systems (Wikipedia, 2011)

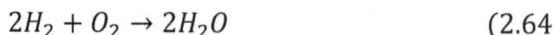

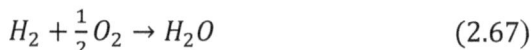

Name of Fuel Cell	Electrolyte	Qualified Power, W	Working Temperature, C	System Efficiency , %
Metal hydride	Aqueous alkaline		➢ -20	

	solution			
Electro-galvanic	Aqueous alkaline solution		< 40	
Direct formic acid (DFAFC)	Polymer membrane	< 50 W	< 40	
Zinc - air	Aqueous alkaline solution		< 40	
regenerative	Polymer membrane		< 50	
Direct Boron Hydride	Aqueous alkaline solution		70	
Alkaline	Aqueous alkaline solution	10 – 100 kW	< 80	62
Direct methanol	Polymer membrane	100 mW – 1kW	90 - 120	10 - 20
Reformed methanol	Polymer membrane	5 – 100 kW	250 - 300	25 - 40
Proton exchange membrane	Polymer membrane	100 W – 500 kW	50 - 120	30 - 50
Phosphoric acid	Molten phosphoric acid	< 10 MW	150 - 200	40; Cogener-ation 90
Molten carbonate	Molten alkaline carbonate	100 MW	600 - 650	47
Tubukar solid oxide (TSOFC)	O^{2-} conducting ceramic oxide	< 100 MW	850 - 1100	55 - 60
Direct carbon			700 - 850	70
Planar solid oxide	O^{2-} conducting ceramic oxide	< 100 MW	500 - 1100	55 - 60
Magnesium - air	Salt water	-20 - 55		

Many considerations are factored into the decision to use co-generation, usually centred around the need for thermal, economic and chemical efficiencies of using power, heat and chemicals. Typical reactions, capable of being run in the generation mode are listed in Table 2.12c.

Vayenas et al (1991), however, found, in their study of cogeneration, in solid oxide fuel cells (SOFCs), that only (i) exothermic reactions with inexpensive materials and (ii) reactions that can be carried out at temperatures higher than 800 C, using existing technology, would be good candidates a for chemical co-generation.

Table 2.12b: Performance Comparisons of Fuel Cells (EERE Information Centre, in Wikipedia, 2011)

Fuel Cell Type	Operating. Temp, C	Stack size	Efficiency
Polymer electrolyte Membrane	50 - 100	<1kW – 100 kW	60 % (transportation); 35 % stationary
Alkaline Fuel Cell	90 – 100	10 – 100 kW	60 %
Phosphoric Acid Fuel Cell	150 – 200	400 kW, 100 kW	40 %
MCFC	600 – 700	300 kW – 3 MW; 300 kW	45 – 50 %
SOFC	700 - 1000	1 kW – 2 MW	60 %

Only few reactions fit this pattern and two examples given were:

a) the oxidation of H_2S to SO_2

$$H_2S + \frac{3}{2}O_2 \rightarrow SO_2 + H_2O \tag{2.70}$$

b) .the partial oxidation of methanol to formaldehyde

$$CH_3OH + \frac{1}{2}O_2 \rightarrow H_2CO + H_2O \tag{2.71}$$

Table 2.12c: Some Co-generative (Electrogenerative) Processes (Langer & Golucci-Rios, 1985)

Overall Reaction	E_0, Theoretical, Volts	E_0, Observed, Volts	Industry
Hydrogenation			
$C_2H_4(g) + H_2(g) = C_2H_6(g)$	0.52	0.51	Ethylene

			Reduction
$C_6H_6(g) + 3H_2(g) = C_6H_{12}(l)$	0.17	0.14	Cyclo-hexane Production
$2NO(g) + H_2(g) = N_2O(g) + H_2O(l)$	1.59	0.90	Pollution Control
Oxidation			
$C_2H_4(g) + 1/2O_2(g) = C_4H_8O$	1.16	n.a.	MEK Production
$SO_2(g) + 1/2O_2 + H_2O(l) = H_2SO_4(aq)$	1.06	0.65	Pollution Control
$H_2 + O_2 = H_2O_2(aq)$	0.68	n.a.	Hydrogen Peroxide
Halogenation			
$C_2H_4(g) + Cl_2(g) = Cl(CH_2)_2Cl(g)$	0.77	0.81	Dichloro-alkane
$C_2H_4(g) + Cl_2(g) + H_2O = Cl(CH_2)_2OH(l)$	0.74	0.81	Chlor-hydrin

By applying what the authors referred to as Non-Faradaic Electrochemical Modification of Catalytic Activity (NEMCA) they were able to postulate that cogeneration may be more useful in enhancing activity and selectivity of catalyst surfaces than in power production. This has implications in promoter selection for conventional supported catalysts or in the use of SOFC-type reactors to carry out reactions where NEMCA can give high selectivity to desired products (Vayenas et al, 1991). There are no industrial units, however, to confirm the expectations of these models.

References for Chapter Two

1. Bailey J. E and Ollis D. F; *Biochemical Engineering Fundamentals*; McGraw-Hill Chemical Engineering Series, McGraw-Hill Book Company, N. Y., USA
2. Bommaraju T. V., Orosz P. J., and E. A. Sokol; *Brine Electrolysis*; Electrochemistry Encyclopaedia; http://electrochemcwru.edu/encycl/
3. *Catalyst Bulletin, 1980*; Harshaw Chemical Company, Ohio, USA

4. *Catalyst Bulletin, 1982*; Spex Industries Corporation, USA
5. Denbigh K. G. & Turner C. R.; *Chemical Reactor Theory*; 2nd Edition, Cambridge University Press, London, 1971
6. Fogler, H Scott, *The Elements of Chemical Kinetics and Reactor Calculations*; Prentice Hall Inc; N. J., USA, 1974
7. Gerrens H; *How to Select Polymerisation Reactors Part II*; CHEMTECH No. 7; pp434 – 442, 1982
8. http://en.wikipedia.org/wiki/ Nuclear_chain_reaction
9. http://en.wikipedia.org/wiki/ fuel cell
10. http://www.ineris.fr/ippc/sites/default/interactive/brefca/bref_gb2.htnl
11. Iqbal, M. Z, 2009; http://polymer.w99of.com/images/polymerisation processes.jpg
12. IUPAC Gold Book definition of elementary reaction
13. Langer, S. H., Card J. C., and M. J. Foral; *Electrogenerative and related processes*; Pure & Appl. Chem., Vol. 58, No. 6, pp. 895 - 906, 1986
14. Langer S. H., and J. A. Colucci-Rios; CHEMTECH, 15 (4), pp 225 – 233, 1985
15. Nnolim B. N.; Unpublished Lecture Notes in Chemical Reaction Engineering; IMT, Enugu, Nigeria; (1989).
16. Novo Industrie, Enzyme Catalog, 1986
17. Vayenas C. G., Bebelis S., and C. Kyriazis; *Solid Electrolytes and Catalysis, Part 1; Chemical Co-generation*; CHEMTECH, 21, 422 – 428 (1991)
18. Vayenas C. G., Bebelis S., and C. Kyriazis; *Solid Electrolytes and Catalysis, Part 2; Non-Faradaic Catalysis*; CHEMTECH, 21, 422 – 428 (1991)

CHAPTER THREE: THE RATE OF CHEMICAL REACTIONS

Example 3.1

Define the rate of a chemical reaction. How would this definition be affected if the phase of the reacting materials is taken into account?

Answer

The rate of a chemical reaction is defined as the rate of formation or disappearance, of a chosen species, per unit time per unit volume, or unit surface or mass, of reaction mixture. Thus if A is the chosen species, n_A = moles of A in the reacting volume while V = volume of the reaction mixture, then, for homogenous systems, the rate of reaction, r_A, is given by

$$-r_A = \frac{1}{V}\frac{dn_A}{dt} \tag{3.1}$$

For heterogeneous systems where the mass, W or surface, S, of reaction mixture, may be more realistic to use, r_A is defined with respect to unit mass or surface and unit time. Thus, based on mass,

$$-r_A^W = \frac{1}{W}\frac{dn_A}{dt} \tag{3.2}$$

and based on surface

$$-r_A^S = \frac{1}{S}\frac{dn_A}{dt} \tag{3.3}$$

For gaseous systems, it is usual to assume ideal gas behaviour such that

$$n_A = \frac{P_A V}{RT} \tag{3.4}$$

so that, if V is constant,

$$-r_A^G = \frac{1}{RT}\frac{dP_A}{dt} \tag{3.5}$$

The rate of reaction, whether defined as $-r_A$, $-r_A^W$, $-r_A^S$ or as r_A^G, is an intensive quantity and represents the intrinsic reaction rate of the reaction, which it describes, on a molecular scale. When the reaction is significantly influenced by transport processes, as in commercial practice, a global rate, based on macro conditions of reaction, is more useful.

The rate of a chemical reaction is a measure of how the concentration or pressure of the substances, involved in the reaction, changes with time.

Example 3.2

Outline the factors which influence the rate of a chemical reaction

Answer

The factors which influence the rate of chemical reaction are:

Reactant concentrations. The higher the concentrations, the faster the rate of reaction because of increased collisions per unit time,

Surface Area. The higher the available surface area, the greater the number of collisions and hence the faster the reaction,

Pressure. Increasing the pressure decreases the volume between molecules thus increasing the frequency of collisions between molecules.

Activation energy, which is the amount of energy required to initiate the reaction and carry it on spontaneously. A reaction that is associated with a high activation energy will require more energy to start than a reaction with a lower activation energy.

Temperature. Higher temperatures increase the energy of the molecules, thereby, creating more collisions per unit time,

The presence or absence of a *catalyst*. Catalysts change the pathway (mechanism) of a reaction by lowering the activation energy thereby increasing the speed of the reaction.

For some reactions, the presence of *electromagnetic radiation*, most notably *ultra violet*, is needed to hasten the breaking of bonds to start the reaction. This is particularly true for reactions involving radicals.

Example 3.3

What is a rate equation with respect to chemical reactions?

Answer

The rate equation is an algebraic equation which shows the form of dependence of the rate of reaction, $-r_A$, on the reacting materials and on the temperature and pressure of the reaction. It is usually applied to reactions taking place on the micro or molecular scale of events and, thus, independent of the type of macro physical reaction system in which the reaction takes place.

Usually, the rate of reaction is a function, at a given total system pressure, of two terms, namely, a temperature dependent term, f(T) and a concentration or partial pressure dependent term, g (C) or g (P). That is

$$Rate\ of\ reaction\ \propto f(T).g(C) \tag{3.6}$$

or

$$Rate\ of\ reaction\ \propto f(T).g(P) \tag{3.7}$$

For the reaction A + B = C, for example, the rate equation may take any of the following forms depending on whether the reaction is elementary or non-elementary,

$$-r_A = k_n[A][B] \ldots. \tag{3.8}$$

$$-r_A = \frac{k_1[A].[B]^n}{k_2 + \frac{[A]}{[B]}} \tag{3.9}$$

or

$$-r_A = \frac{k_1[A]}{1 + k_2[A]} \tag{3.10}$$

where the k_i represent the temperature dependent function, *f (T)* and the various [A], [B], etc, represent the concentration dependent functions.

The chemical reaction rate equation is not, necessarily, a differential equation as would be implied by the fact that $-r_A = \frac{1}{V}\frac{dn_A}{dt}$ looks like one.

Example 3.4

What is the order of a chemical reaction?

Answer

If the rate equation of an elementary, chemical, reaction can be expressed, at constant temperature, as a function of concentration raised to a power, the power to which concentration is raised is called the order of the reaction.

Such a reaction can be described as zero, first, second or third order reaction, depending on the value of the exponent to which the concentration is raised.

In such cases, the rate of reaction, - r_A is given as

$$-r_A = k_n C_A^n \tag{3.11}$$

Such a reaction is said to be an nth order reaction. k_n is the reaction rate constant of the nth order reaction, dependent only on temperature. Thus if

n = 0	$-r_A = k_0$	the reaction is zero order with respect to A
n = 1	$-r_A = k_1 C_A$	the reaction is first order with respect to A
n = 2	$-r_A = k_2 C_A^2$	the reaction is second order with respect to A
n = 3	$-r_A = k_3 C_A^3$	the reaction is third order with respect to A.

Rarely, does n exceed 3. Variations do occur when n = 2 or 3.

For example, when n = 2, and reactants are A and B, the rate equation may, also, take the form

$$-r_A = k_2 C_A C_B \tag{3.12}$$

Similarly, when n = 3 and the reactants are A, B and C, - r_A may, also, take the forms

$$-r_A = k_3 C_A C_B C_C \tag{3.13}$$

or

$$-r_A = k_3 C_A C_B^2 \tag{3.14}$$

or

$$-r_A = k_3 C_A^2 C_B \tag{3.15}$$

etc.

For complex reactions, however, the concept of reaction order, is not, often, useful as the final rate equation can be a combination of several rate equations of participating elementary reactions. The resulting equations may take various forms, such as the ones shown below, or even more complex ones.

$$-r_A = \mathrm{k}[\mathrm{C_A}]^{0.5}[\mathrm{C_B}]^{1.5} \tag{3.16}$$

$$-r_A = \frac{k_1 [A].[B]^n}{k_2 + \frac{[A]}{[B]}} \tag{3.17}$$

or

$$-\mathrm{r_A} = \frac{\mathrm{k_1[A]}}{1+\mathrm{k_2[A]}} \tag{3.18}$$

Complex reactions are better classified by mechanism of reaction such as whether the reactions are parallel reactions, consecutive reactions, chain reactions, clock reactions, fermentation reactions, polymerisation reactions, etc.

Example 3.5

The rate equation for a chemical reaction is the product of a temperature dependent term and a concentration dependent term. Explain the nature of the temperature dependent term.

Answer

A comparison of equations (3.11) to (3.18) would indicate that, k, the reaction rate constant, is the temperature dependent part of the reaction rate equation. That is

$$k = f(T) \tag{3.19}$$

The exact nature of this functional dependence of k on temperature is based on two postulates which attempt to explain the manner in which chemical reactions occur.

The first and earlier of the two postulates is the *Collision Theory* of chemical reactions. The second and later postulate is the *Transition State Theory* of chemical reactions.

Collision Theory of Chemical Reactions

In *collision theory* of gas reactions, chemical reaction is assumed to take place because molecules of reactants approach each other so closely that they can be said to have collided with each other. These collisions then result in chemical reaction. The sequence is:

- Molecules approach each other
- Reaction occurs when their distance is equal to their collision diameter

The speed of reaction is then the product of the *number of collisions per second* and the *fraction of collision which are effective* in producing chemical change.

Collision theory is closely related to chemical kinetics and is derived from the kinetic theory of gases. Reaction rate tends to increase with concentration - a phenomenon explained, also, by the collision theory.

Transition State Theory of Chemical Reactions

This is a more detailed postulate which attempts to explain, better than the collision theory, the manner in which chemical reactions occur. The transition state theory is also known as activated-complex theory or theory of absolute reaction rates.

In the theory, the reactants combine to form unstable intermediates called activated complexes.

$$A + B \rightarrow Activated\ Complex \rightarrow Products \qquad (3.20)$$

These activated complexes then decompose spontaneously to form products. The theory further assumes that, at all times, reactants and complexes are in equilibrium and that the rate of decomposition of complex is the same for all reactions.

Example 3.6

Explain and contrast the predictions of the reaction rate constant by the collision and transition state theories.

Answer

The rate constant as predicted by the collision theory is based on the following assumptions

1. for bimolecular collisions of like molecules, A and A, the kinetic theory gives the number of molecular collisions per second per cubic centimetre, Z_{AA} as

$$Z_{AA} = \sigma_A^2 n_A^2 \sqrt{\frac{4\pi k_B T}{M_A}} = \sigma_A^2 \left(\frac{N^2}{10^6}\right)\left[\sqrt{\frac{4\pi k_B T}{M_A}}\right] . C_A^2 \qquad (3.21)$$

where, for the A molecules,

σ_A = diameter of molecules, cm
M_A = mass of molecule
= molecular weight/Avogadro's number
N = Avogadro's number = 6.023 x 10^{23}, molecules/gmole

C_A = concentration, g/l
n_A = number of molecules per cc = $10^{-3}N.C_A$
k_B = Boltzman's constant = R/N
R = universal gas constant

2. for bimolecular collisions of unlike molecules, A and B, the kinetic theory gives the number of molecular collisions per second per cubic centimetre, Z_{AB} as

$$Z_{AB} = \left(\frac{\sigma_A + \sigma_B}{2}\right)^2 \left(\frac{N^2}{10^6}\right)\left[\sqrt{8\pi k_B T\left(\frac{1}{M_A} + \frac{1}{M_B}\right)}\right] C_A C_B \quad (3.22)$$

By defining

$$\sigma_{AB} = \left(\frac{\sigma_A + \sigma_B}{2}\right) = reaction\ cross\ section \quad (3.23)$$

$$\mu_{AB} = \left(\frac{1}{\frac{1}{M_A} + \frac{1}{M_B}}\right) = reduced\ mass\ of\ reactants \quad (3.24)$$

equation (3.22) is found to reduce to

$$Z_{AB} = \sigma_{AB}^2 \left(\frac{N^2}{10^6}\right)\left[\sqrt{\frac{8\pi k_B T}{\mu_{AB}}}\right] . C_A . C_B \quad (3.25)$$

Equation (3.25) may, also, be expressed as

$$Z_{AB} = Z_0\ . C_A . C_B \quad (3.26)$$

where

$$Z_0 = \sigma_{AB}^2 \left(\frac{N^2}{10^6}\right)\left[\sqrt{\frac{8\pi k_B T}{\mu_{AB}}}\right] \quad (3.27)$$

3. the fraction of collisions which are effective in producing chemical reaction is given by the Arrhenius theory, as

$$f_E = e^{-\frac{\Delta E}{RT}} \quad (3.28)$$

where ΔE is the minimum energy required to effect a reaction, (the activation energy), R, the universal gas constant and T the temperature of reaction.

$$Speed\ of\ reaction\ = Z_{AB}f_E = Z_0\ .e^{-\frac{\Delta E}{RT}}.C_A.C_B,\quad molecular\ collisions\ per\ second\ per\ cm^3 \tag{3.29}$$

When converted to gmoles per litre per second, we get

$Speed\ of\ reaction$

$$= Z_0\ .e^{-\frac{\Delta E}{RT}}.C_A.C_B\left\{\frac{molecular\ collisions}{cm^3\ .\ second}\right\}\left(\frac{1000}{N}\right)\left\{\frac{cm^3}{litre}\right\}x\left\{\frac{gmole}{molecules}\right\}$$

$$= Z_0 C_A C_B\ e^{-\frac{\Delta E}{RT}}.\frac{1000}{N},\frac{gmole}{litre\ .\ second} \tag{3.30}$$

But

$$Speed\ of\ reaction = -r_A = \frac{1}{V}\frac{dn_A}{dt} = k_2 C_A C_B \tag{3.31}$$

Hence, from equations (3.30) and (3.31)

$$k_2 C_A C_B\ = Z_0 C_A C_B\ e^{-\frac{\Delta E}{RT}}.\frac{1000}{N}$$

from which

$$k_2 = Z_0\ e^{-\frac{\Delta E}{RT}}.\frac{1000}{N} \tag{3.32}$$

Equation (3.32) is the result of theoretical analysis. In practice, the k_2 obtained theoretically is multiplied by a steric factor, ρ, which is defined as the ratio between the experimental value of the rate constant and that predicted by collision theory. It is most often less than unity although, some reactions (harpoon reactions) exhibit steric factors greater than unity.

The simple collision theory, summarised above, does not give a clear interpretation of the activation energy, nor a way to theoretically calculate it, but, in spite of its simplicity, it provides a basis for defining typical kinetic behaviour, which allows us to focus on more particular cases.

The transition state theory of chemical reactions is a more detailed postulate which attempts to explain, better than the collision theory, the manner in which chemical reactions occur. The transition state theory is also known as activated-complex theory or theory of absolute reaction rates.

The transition state theory assumes that once a reaction passes through its reaction barrier it cannot go back again. Whilst transition state theory is a large improvement on simple collision theory, it requires knowledge of the partition function of intermediate species which, due to their high energy and short lifespan, can be hard to determine spectroscopically.

The theory further assumes that, at all times, reactants and complexes are in equilibrium and that the rate of decomposition of complex is the same for all reactions and is given by k_BT/h. where k_B is the Boltzman constant, h is the Planck constant = 6.6262×10^{-34} Joule-sec and $k_B = R/N = 1.3806 \times 10^{-23}$ Joules/deg K. The rate of reaction is then

$$r_A = \frac{k_BT}{h}\ x\ concentration\ of\ activated\ complex \quad (3.33)$$

For example, the proposed reaction sequence for the reaction

$$A + B \rightleftharpoons AB \quad (3.34)$$

would, according to the transition state theory, be

$$A + B \rightleftharpoons (AB)^* \rightarrow AB \quad (3.35)$$

The concentration based equilibrium constant for the formation of the activated complex is given by:

$$K_C^* = \frac{C_{(AB)^*}}{C_AC_B} \quad (3.36)$$

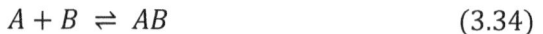

from which

$$C_{(AB)^*} = K_C^* C_A C_B \tag{3.37}$$

Substituting (3.37) into (3.33), we get

$$r_A = \frac{k_B T}{h} \cdot C_{(AB)^*} = \frac{k_B T}{h} \cdot K_C^* \,.\, C_A \, C_B \tag{3.38}$$

K_C^* is calculated, from thermodynamics, to be $e^{\left(-\frac{\Delta H'}{RT}+\frac{\Delta S'}{R}\right)}$ so that equation (3.38) becomes

$$r_A = \frac{k_B T}{h} \cdot e^{\frac{\Delta S'}{R}} \cdot e^{-\frac{\Delta H'}{RT}} \,.\, C_A \, C_B \tag{3.39}$$

As with the collision theory, to relate theoretical calculations to experimental results, equation (3.39) is multiplied by an arbitrary factor, φ, which has a value between 0 and 1. Thus

$$r_A = \varphi \frac{k_B T}{h} \cdot e^{\frac{\Delta S'}{R}} \cdot e^{-\frac{\Delta H'}{RT}} \,.\, C_A \, C_B \tag{3.40}$$

The reaction rate constant can, now, be seen to be

$$k = \varphi \frac{k_B T}{h} \cdot e^{\frac{\Delta S'}{R}} \cdot e^{-\frac{\Delta H'}{RT}} \tag{3.41}$$

Equation (3.40) is, usually, approximated as

$$r_A = A \,.\, T \,.\, e^{-\frac{\Delta E}{RT}} \,.\, C_A \, C_B \tag{3.42}$$

where A is a constant equal to $\varphi \,.\frac{k_B}{h} \,.\; e^{\frac{\Delta S'}{R}}$ and ΔE is an activation energy

Example 3.7

Explain the term: Activation Energy of Chemical Reactions

Answer

The activation energy of a chemical reaction is the minimum amount of energy required for the reaction to occur. Both the collision theory and the transition state theory of chemical reactions accept that an activation energy is necessary for a chemical reaction to proceed. Figure 3.7 below illustrates graphically how this energy changes as the reaction proceeds.

Position A, on the diagram, represents the energy level of the reactants at the beginning of the reaction. Position B represents the energy level of the activated complex while position C represents the energy level of the products of the reaction. ΔE is used, instead of just E, to remind you that these energy levels are energy changes or differences with respect to a reference state

Fig. 3.7 illustrates, also, the relationship between activation energy, ΔE, and enthalpy of formation, ΔH.

Fig. 3.7: Illustration of the Activation Energy of Chemical Reactions

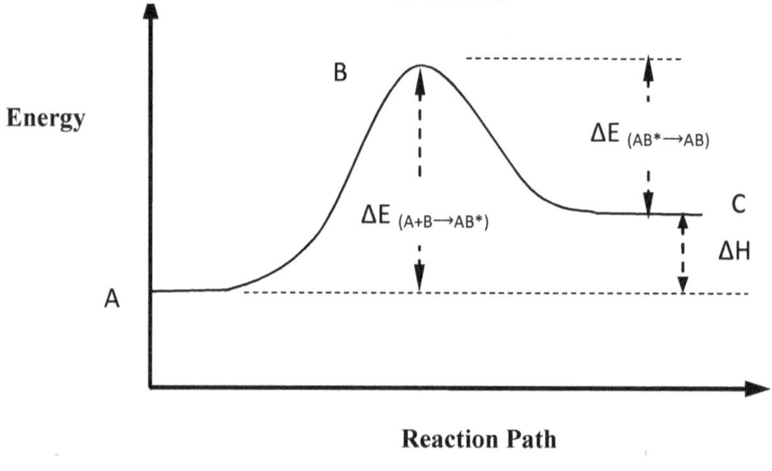

The peak energy position represents the transition state. Svante Arrhenius found out, experimentally, (Moore J), that

$$k = f(T) = k_0 e^{-\frac{\Delta E_A}{RT}} \qquad (3.43)$$

Equation (3.43) is known as the Arrhenius equation and ΔE_A, the activation energy for the reaction. k_0 is called the frequency factor by comparison with results derived from the collision theory.

For a general reversible reaction such as

$$A \underset{k_2}{\overset{k_1}{\rightleftharpoons}} B \tag{3.44}$$

the van't Hoff equation is

$$\frac{dlnK}{dT} = \frac{\Delta H_R}{RT^2} \tag{3.45}$$

where ΔH_R is the heat of reaction and K, the equilibrium constant for the reaction, equal to k_1/k_2. Equation (3.45) can be written as

$$\frac{dlnK}{dT} = \frac{dln\, k_1}{dT} - \frac{dln\, k_2}{dT} = \frac{\Delta H_R}{RT^2} \tag{3.46}$$

This implies that

$$\frac{\Delta H_R}{RT^2} = \frac{\Delta E_1}{RT^2} - \frac{\Delta E_2}{RT^2} \tag{3.47}$$

where ΔE_1 and ΔE_2 are the activation energies for the forward and backward reactions. High activation energies result in reactions that are very sensitive to temperature changes and vice versa.

Example 3.8

Illustrate, with examples, the manner in which activation energy affects the rate of chemical reactions.

Answer

Although the rate of reactions depends on concentration, temperature and pressure, the factor that is most affected by the activation energy is the reaction rate constant. This constant, in turn, is dependent on the temperature of reaction.

At high temperatures, the reaction rate constant is, generally, higher regardless of the level of activation energy than at low temperatures. What is affected by the activation energy is the rate at which the reaction rate constant varies with temperature.

Fig 3.8: Effect of Activation Energy on Reaction Rate Constant

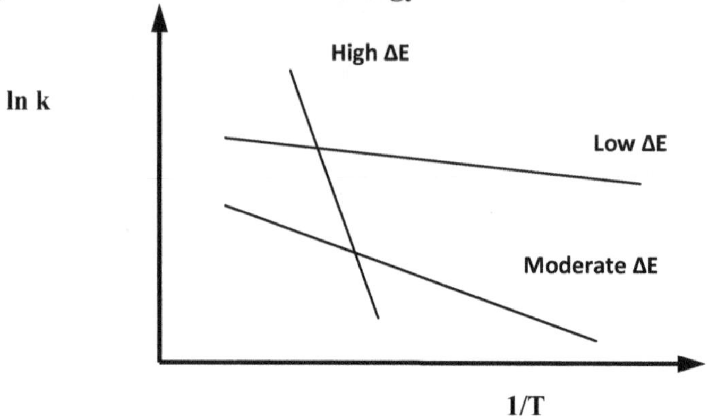

At high activation energies, the reaction rate constant decreases rapidly with increase in reaction temperature. At low activation energy levels, this decrease is quite gradual while at moderate activation energies, the decrease is somewhere between that of high and low activation energies. These effects are summarised as shown in Fig. 3.8.

Example 3.9

Outline, briefly, the reported observed effects of temperature on the rates of chemical reactions

Answer

Temperature has enormous effect temperature on the rate of reactions. These effects have been summarised by Walas and are summarised in Fig. 3.9.I to 3.9.VI.

Fig. 3.9: Observed Effects of Temperature on Reaction Rate (Walas, 1959)

I: Normal Effect – Rapid Increase of Reaction Rate with Temperature

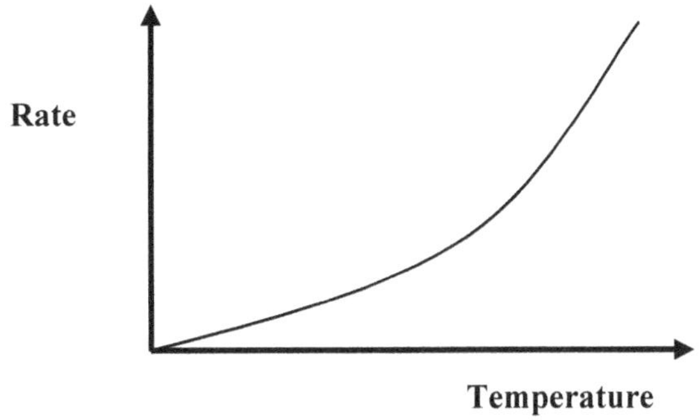

II: Heterogenous Reactions – Diffusion Controls (Moderate Temperature Effect)

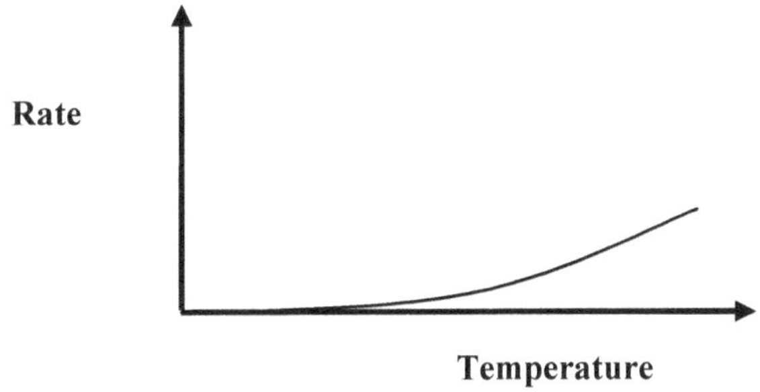

III: Explosive Reactions

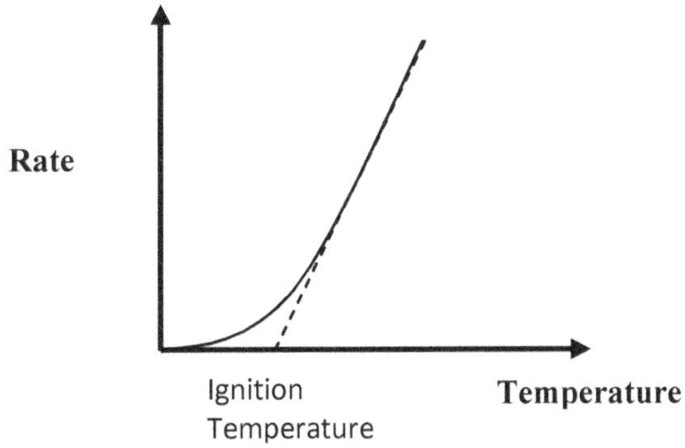

IV: Catalytic Reactions (Rate of adsorption controls and Decreases at High Temperature) and Enzyme Reactions (High Temperature Deactivates Enzyme)

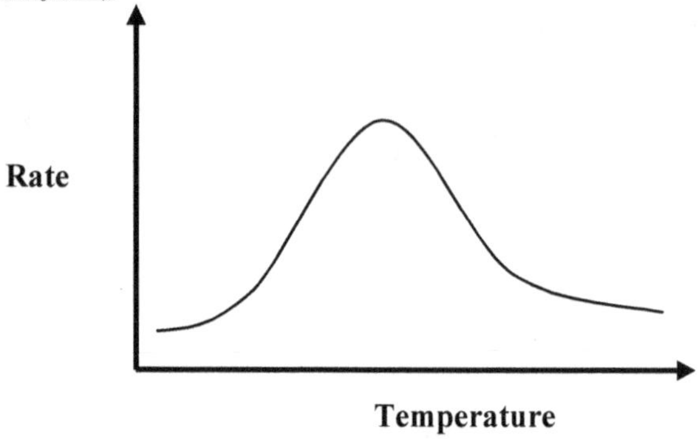

V: Reactions Compilcated by Side Reactions which become significant at High Temperatures

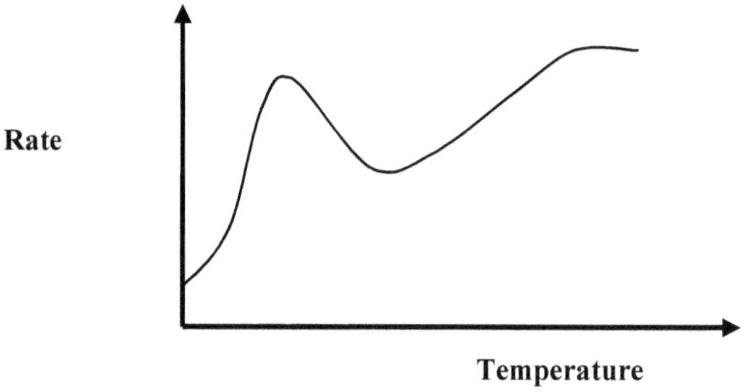

VI: Reactions Favoured by Lower Temperatures

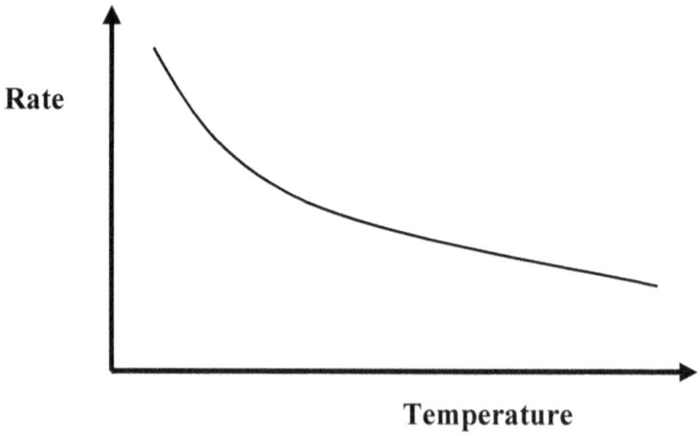

Example 3.10

It is said that a catalysed reaction proceeds faster than an uncatalysed reaction, by a mechanism involving lower activation energy. Explain

Answer

A catalyst is able to make a catalysed reaction proceed faster than an uncatalysed reaction by lowering the activation energy of the catalysed reaction. How this is done may be seen by looking at the potential energy map of the reaction as shown in Figs. 3.10a and 3.10b.

Fig. 3.10a shows the potential energy of the reactants at their initial state, the potential energy of the transition state or activated complex and the potential energy of the products or the final state for the uncatalysed reaction. From this can be seen the relative values and position of the activation energy and the heat of reaction.

Fig. 3.10a: Potential Energy Map of Uncatalysed Reaction

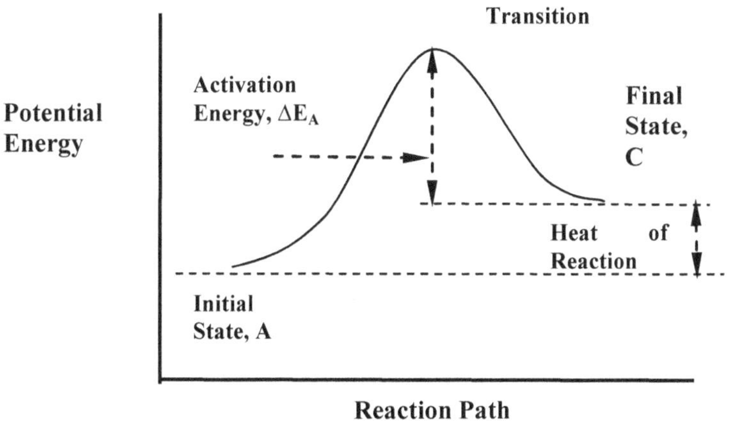

Fig. 3.10b shows the same type of diagram, overlaid on a similar one for a catalysed reaction. It shows the formation of an intermediate compound at *B* is associated with less activation

energy than that of the uncatalysed reaction. Note that, for the same reaction, the heat of reaction is still the same for both the catalysed and uncatalysed reaction.

Fig. 3.10b: Potential Energy Map of the Catalysed Reaction

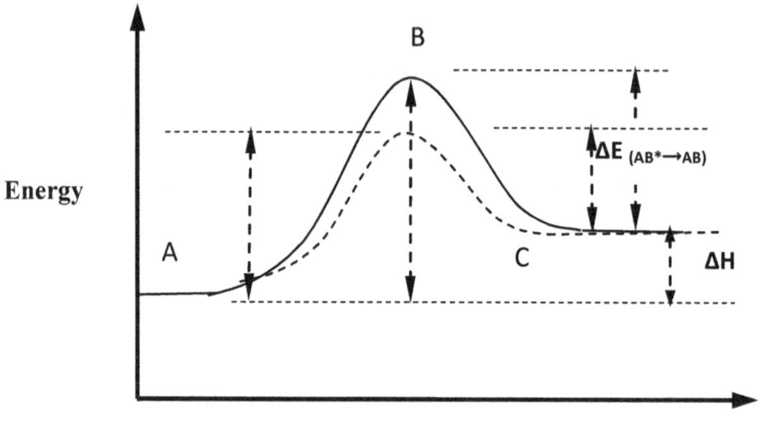

Example 3.11

List, with very brief explanations, the various methods by which the rate expression of a chemical reaction may be obtained in the laboratory.

Answer

The rate expression of a chemical reaction may be obtained, in the laboratory, by any of the following methods

a. <u>Methods based on measuring the concentrations of reactants and/or products</u>

Differential analysis	The rate constants are obtained using differential changes in concentration, dC, within specified time and concentration limits
Integral method	Rate constants are obtained by integration of applicable equations over given or selected

	concentration or time limits
Method of critical rates	The rate constants are based on the rate determining step or concentration
Method of half lives	The rate constants are based on when the concentration of some key reactant reaches half its initial value

b. Methods based on the temperature dependent terms of the rate expression

In these methods, the reaction rate constant, *k*, is determined at three or more different temperatures, keeping the concentration dependent terms the same throughout. A plot of the Arrhenius equation will yield the average values of *k* and ΔE within the interval of temperatures used.

Example 3.12

State, briefly, the steps required to arrive at a rate expression for a chemical reaction.

Answer

The first step is to write the chemical equation of the reaction and balance it.

The second step is to determine, from the nature of the balanced equation and any other previous experience, whether the reaction can be considered an elementary reaction or not.

The third step is to determine what the physical conditions of the reaction are; whether it occurs at constant pressure or temperature, whether it is a homogenous or heterogeneous reaction, whether it seems reversible or not, etc.

The fourth step is to propose a mechanism for the reaction based on the above considerations

The fifth step is to develop a rate expression based on the proposed mechanism of the reaction.

The sixth step is to test the proposed mechanism and rate expression using the results of experimental measurements.

Example 3.13

Given a reaction of known order which also fits the collision theory model of chemical reaction, describe, briefly but clearly, how you would set out, experimentally, to determine the rate constant of the reaction.

Answer

For a reaction of known order, *n*, which fits the collision theory model of chemical reaction, the rate equation, based on key reactant A, at constant reactor volume, is given by

$$-r_A = \frac{dC_A}{dt} = kC_A{}^n \qquad (1)$$

The first step is to obtain, at a known constant temperature and total system pressure, experimental values of C_A as a function of time, t, during chemical reaction. Care should be taken to identify sources and magnitudes of errors associated with each measurement. In many cases, to avoid bias in the measurements, the sequence of readings is randomised

The next step is to plot a graph of C_A versus t or set up a table if numerical differentiation is to be used.

Next step is to obtain values of $\frac{dC_A}{dt}$ at various values of C_A , graphically or numerically.

Using either a log-linear or linear – linear graph paper, a plot of $\log\left(\frac{dC_A}{dt}\right)$ versus $\log C_A$ is made since, from equation (1)

$$\log\left(\frac{dC_A}{dt}\right) = \log k + n\, logC_A. \qquad (2)$$

This plot usually results in a straight line having a slope equal to n and an intercept on the y-axis equal to $\log k$. Alternatively, polynomial regression methods could be used to obtain k and n

Example 3.14

Chlorine gas is produced in a zero order reaction between carbon tetrachloride (CCl_4) and water in an ultrasonic field. Data obtained is as follows:

Time, min	0	1	2	3	4
Concentration of Chlorine, meq/l	0.000	0.090	0.170	0.265	0.385

What is the rate constant for this reaction?

Answer

For a zero order reaction

$$\frac{dC_{Cl_2}}{dt} = k_o$$

Thus

$$\int_{C^o_{Cl_2}}^{C_{Cl_2}} dC_{Cl_2} = \int_0^t k_o\, dt$$

from which

$$C_{Cl_2} - C^o_{Cl_2} = k_o t \qquad (1)$$

A graphical plot of C_{Cl_2} versus t should give a straight line of slope k_o. This graphical plot is shown Figure 3.14 below.

The slope of the best straight line through the data is given by

$$k_o = \frac{C^t_{Cl_2} - C^o_{Cl_2}}{t - 0} = \frac{0.360 - 0}{4 - 0} = 0.09\, mEq/litre.min$$

Fig. 3.14: Concentration of Chlorine versus Time

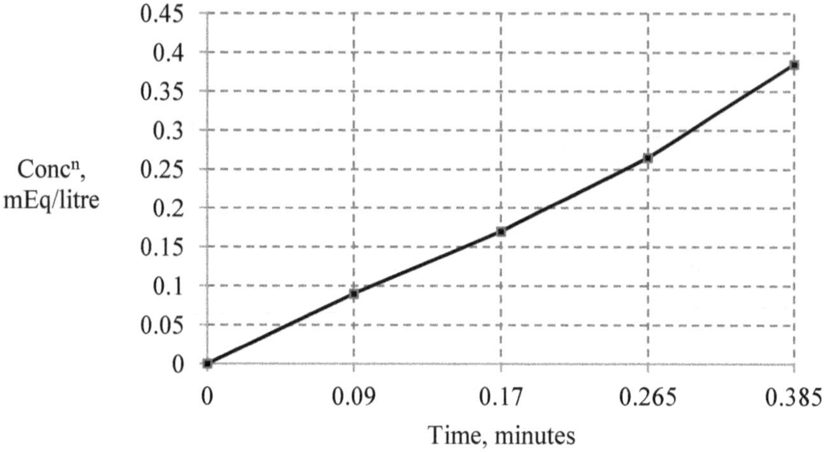

The linear regression equivalent equation is

$$C_{Cl_2} = C^o_{Cl_2} + k_o t \;\; = y = a + bt \qquad (2)$$

With linear regression, the Table below may be constructed

	t, min	$y = C_{Cl_2}, mEq/litre$	y_t, **mEq min/litre**	t^2
	0	0	0	0
	1	0.090	0.090	1
	2	0.170	0.340	4
	3	0.265	0.795	9
	4	0.385	1.54	16
Totals	10	0.910	2.765	30

$$a = \frac{\sum_1^4 y \sum_1^4 t^2 - \sum_1^4 t \sum_1^4 ty}{4\sum_1^4 t^2 - (\sum_1^4 t)^2} = \frac{0.91\ x\ 30 - 10\ x 2.765}{4\ x\ 30 - 10\ x\ 10}$$

$$= -0.0175 \qquad (3)$$

$$b = \frac{4\sum_1^4 ty - \sum_1^4 t \sum_1^4 y}{4\sum_1^4 t^2 - (\sum_1^4 t)^2} = \frac{4\,x\,2.765 - 10\,x\,0.91}{4\,x\,30 - 10\,x\,10}$$

$$= 0.098 \qquad (4)$$

Note that this analysis gives $a = C_{Cl_2}^{o}$ as -0.0175 instead of 0 and $b = k_o = 0.098$.

Example 3.15

The extent of decomposition of vinyl ethyl ether at 389 C at various times, for an initial pressure of 51 torr, is given as

% Decomposition	20	30	40	50
Time, seconds	264	424	609	820

Determine the rate constant for the reaction.

Answer

If we represent vinyl ethyl ether by A, the reaction could be represented as

$$A \rightarrow Products$$

Let us assume a constant volume reactor, and in the absence of any information to the contrary, that the reaction is first order. Then

$$-r_A = -\frac{dC_A}{dt} = k_1 C_A$$

so that

$$\frac{C_{Ao} dX_A}{dt} = k_1 C_{Ao}(1 - X_A)$$

That is

$$\int \frac{dX_A}{1 - X_A} = k_1 t$$

from which we have that

$$-\ln(1 - X_A) = k_1 t \qquad (1)$$

A graphical plot of $-\ln(1 - X_A)$ versus t will have a slope equal to k_1. The table of values is calculated as shown in the Table below.

X_A	$1 - X_A$	$-\ln(1 - X_A)$	t, s
0.2	0.8	0.223	264
0.3	0.7	0.357	424
0.4	0.6	0.511	609
0.5	0.5	0.693	820

The graphical plot is shown in Fig. 3.15. The slope of the best fitting straight line is

$$slope = k_1 = \frac{0.690 - 0.200}{820 - 240} = 8.448\ x\ 10^{-4}, sec^{-1}$$

This confirms that the reaction is first order. Ans

Fig. 3.15: Plot of -ln (1-X_A) versus Time

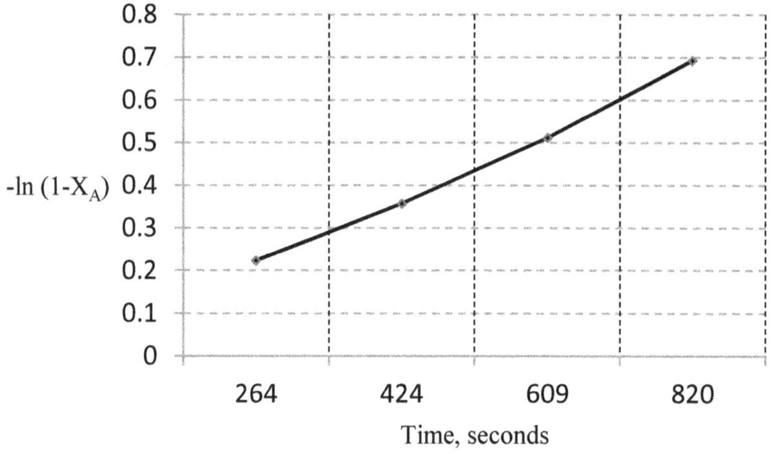

Example 3.16

The reaction

$$SO_2Cl_2 \rightarrow SO_2 + Cl_2$$

is a first order reaction with $k_1 = 2.2 \times 10^{-5}$ sec^{-1} at 320 C. What percent of SO_2Cl_2 is decomposed on heating at 320 C for 90 minutes?

Answer

Let C_A be the concentration of SO_2Cl_2 at time t.

Since the reaction is first order

$$\frac{C_A}{C_{Ao}} = e^{-k_1 t}$$

$$k_1 t = 2.2 \; x \; 10^{-5} \; x \; 90 \; x \; 60 = 0.1188$$

Hence

$$\frac{C_A}{C_{Ao}} = e^{-0.1188} = 0.888$$

$$Conversion = X_A = 1 - \frac{C_A}{C_{Ao}} = 1 - 0.888 = 0.112$$

$$Percent\ decomposition = 11.2\ \% \quad Ans$$

Alternatively,

$$X_A = 1 - \frac{C_A}{C_{Ao}} = 1 - e^{-k_1 t} = 1 - e^{-0.1188} = 0.112$$

$$Percent\ decomposition = 11.2\ \% \quad Ans$$

Example 3.17

Derive the user friendly expression for the rate equation of a second order, elementary, irreversible reaction of the type

$$2A \rightarrow Products$$

Answer

For this reaction

$$-r_A = -\frac{dC_A}{dt} = k_2 {C_A}^2$$

Thus

$$\int_{C_{Ao}}^{C_A} \frac{dC_A}{C_A{}^2} = -\int_0^t k_2\, dt$$

That is

$$\left| -\frac{1}{C_A} \right|_{C_{Ao}}^{C_A} = -k_2 t$$

From which

$$\frac{1}{C_A} - \frac{1}{C_{Ao}} = k_2 t \qquad Ans$$

This expression relates the initial concentration to the concentration at any time, t, and to the second order rate constant.

In terms of fractional conversion

$$-r_A = \frac{C_{Ao} dX_A}{dt} = k_2 C_{Ao}{}^2 (1 - X_A)^2$$

That is

$$\int_0^{X_A} \frac{dX_A}{(1 - X_A)^2} = -\int_0^t k_2\, C_{Ao} dt$$

Thus

$$\frac{X_A}{1 - X_A} = C_{Ao} k_2 t \qquad Ans$$

Example 3.18

Derive the user friendly expression for the rate equation for a second order, elementary, irreversible, reaction of the type

$$A + B \rightarrow Products$$

Answer

$$-r_A = -\frac{dC_A}{dt} = k_2 C_A C_B$$

It is easier to analyse this in terms of fractional conversion. Thus

$$\frac{C_{Ao}dX_A}{dt} = k_2 C_{Ao}(1 - X_A)C_{Bo}(1 - X_B)$$

But

$$C_{Ao}X_A = C_{Bo}X_B$$

That is

$$X_B = \frac{C_{Ao}}{C_{Bo}}X_A$$

Hence

$$\frac{C_{Ao}dX_A}{dt} = k_2 C_{Ao}(1 - X_A)C_{Bo}\left(1 - \frac{C_{Ao}}{C_{Bo}}X_A\right)$$

If we define

$$\theta = \frac{C_{Bo}}{C_{Ao}} = initial\ molar\ concentration\ ratio$$

Then

$$\frac{C_{Ao}dX_A}{dt} = k_2 {C_{Ao}}^2(1 - X_A)(\theta - X_A)$$

from which

$$\int_0^{X_A} \frac{dX_A}{(1 - X_A)(\theta - X_A)} = \int_0^t k_2\, C_{Ao}dt$$

From the mathematics of partial fractions

$$\frac{1}{(1 - X_A)(\theta - X_A)} = \frac{1}{(\theta - 1)(1 - X_A)} - \frac{1}{(\theta - 1)(\theta - X_A)}$$

Thus

$$\int_0^{X_A} \frac{dX_A}{(1 - X_A)(\theta - X_A)} = \frac{1}{(\theta - 1)}\int_0^{X_A} \frac{dX_A}{(1 - X_A)} - \frac{1}{(\theta - 1)}\int_0^{X_A} \frac{dX_A}{(\theta - X_A)}$$

$$= -\frac{1}{(\theta - 1)}\ln(1 - X_A) + \frac{1}{(\theta - 1)}\ln\frac{(\theta - X_A)}{\theta}$$

That is

$$\frac{1}{(\theta - 1)}\ln\frac{(\theta - X_A)}{\theta(1 - X_A)} = C_{Ao}k_2 t \qquad Ans$$

or

$$\frac{(\theta - X_A)}{\theta(1 - X_A)} = \exp\big((\theta - 1)C_{Ao}k_2 t\big) \qquad Ans$$

To express this result in terms of concentrations, note that

$$\theta = \frac{C_{Bo}}{C_{Ao}}.$$

and

$$X_B = \frac{C_{Ao}}{C_{Bo}} X_A$$

Hence

$$\frac{(\theta - X_A)}{\theta(1 - X_A)} = \frac{(C_{Bo} - X_A C_{Ao})}{C_{Ao}(1 - X_A)} \cdot \frac{C_{Ao}}{C_{Bo}} = \frac{(C_{Bo} - X_B C_{Bo})}{C_{Ao}(1 - X_A)} \cdot \frac{C_{Ao}}{C_{Bo}} = \frac{C_B}{C_A} \frac{C_{Ao}}{C_{Bo}}$$

From which we get that

$$\frac{(\theta - X_A)}{\theta(1 - X_A)} = \frac{C_B}{C_A} \frac{C_{Ao}}{C_{Bo}} = \exp\left((C_{Bo} - C_{Ao}) k_2 t\right) \qquad Ans$$

Example 3.19

The hydrolysis of ethyl nitro benzoate by hydroxyl ions

$$NO_2C_6H_5COOC_2H_5 + OH^- \rightarrow NO_2C_6H_4COOH + C_2H_5OH$$

proceeds as follows at 15 C (J. Chem. Soc (1936) in Moore (1969)) when the initial concentrations of both reactants are 0.05 mole per litre

Time, s	120	180	240	330	530	600
% hydrolysed	32.95	41.75	48.8	58.05	69.0	70.33

Calculate the second order rate constant

Answer

Since θ = 1, we cannot use

$$\frac{1}{(\theta - 1)} \ln \frac{(\theta - X_A)}{\theta(1 - X_A)} = C_{Ao} k_2 t$$

But we can use

$$\frac{X_A}{1-X_A} = C_{Ao}k_2t$$

From which

$$k_2 = \frac{X_A}{C_{Ao}(1-X_A)t}$$

At t = 120 s and X_A = 0.3295

$$k_2 = \frac{0.3295}{0.05(1-0.3295)\ x\ 120}, \frac{litre}{moles}\frac{1}{s} = 0.0819\ \frac{litre}{moles.sec}$$

Similarly at the other times and conversions k_2, is obtained as shown below

Time, s	$k_2, \frac{litre}{moles.sec}$
120	0.0819
180	0.0796
240	0.0794
330	0.0839
530	0.0840
600	0.0790
Average	0.0813

Hence

$$k_2 = 0.0813\ \frac{litre}{moles.sec} \qquad Ans$$

Example 3.20

The half life of a chemical reaction is defined as the time in which the concentration of the key reactant reaches half its initial value.

Derive the expressions for the half lives of the zero, first and second order elementary, irreversible, chemical, reactions and comment on the results.

Answer

The variation of concentration with time for a zero, first and second order elementary, irreversible chemical reactions are as shown below. Let $\tau_{½}$ denote the half life of the reaction

For a <u>zero order reaction</u>

$$C_{Ao} - C_A = k_o t$$

When t = $\tau_{½}$, $C_A = ½C_{Ao}$

$$C_{Ao} - ½C_{Ao} = ½C_{Ao} = k_o\tau_{½}$$

from which

$$\tau_{½} = \frac{C_{Ao}}{2k_o} \quad Ans$$

For a <u>first order reaction</u>

$$\frac{C_A}{C_{Ao}} = e^{-k_1 t}$$

When t = $\tau_{½}$, $C_A = ½C_{Ao}$

$$\frac{½C_{Ao}}{C_{Ao}} = \frac{1}{2} = e^{-k_1\tau_{½}}$$

from which

$$\tau_{½} = \frac{\ln 2}{k_1} \quad Ans$$

For a <u>second order reaction</u>

$$\frac{1}{C_A} - \frac{1}{C_{Ao}} = k_2 t$$

When t = $\tau_{½}$, $C_A = ½C_{Ao}$

$$\frac{2}{C_{Ao}} - \frac{1}{C_{Ao}} = \frac{1}{C_{Ao}} = k_2\tau_{½}$$

from which

$$\tau_{½} = \frac{1}{k_2 C_{Ao}} \quad Ans$$

For an nth order elementary reaction

$$\frac{C_A{}^{1-n} - C_{Ao}{}^{1-n}}{(n-1)} = k_n t$$

When t = $\tau_{½}$, $C_A = ½C_{Ao}$

$$\frac{(½C_{Ao})^{1-n} - C_{Ao}{}^{1-n}}{(n-1)} = \frac{2^{n-1}-1}{(n-1)} C_{Ao}{}^{1-n} = k_n \tau_{½}$$

from which

$$\tau_{½} = \frac{2^{n-1}-1}{k_n(n-1)} C_{Ao}{}^{1-n} \quad Ans$$

These results show that the half life of a first order reaction is independent of concentration, that of a zero order reaction directly proportional to the initial concentration while those of the second and higher order reactions are inversely proportional to the initial concentration.

Example 3.21

Find the rate equation for the aqueous reaction which is given by

$$A = B + C$$

Experiments gave the following data

Time, minutes	C_A, mol/litre
0	0.1823
30	0.1515
60	0.1257
90	0.108
120	0.0948
150	0.0833

and C_{Ao} = 0.1823 mole/litre, C_{Bo} = 0, C_{Co} = 0 mole/litre

Answer

The first thing is to plot the data of C_A versus t because its shape will be a good guide as to the order of the reaction. This plot is shown in Fig. 3.21a below.

Fig. 3.21a: Plot of C_A versus Time

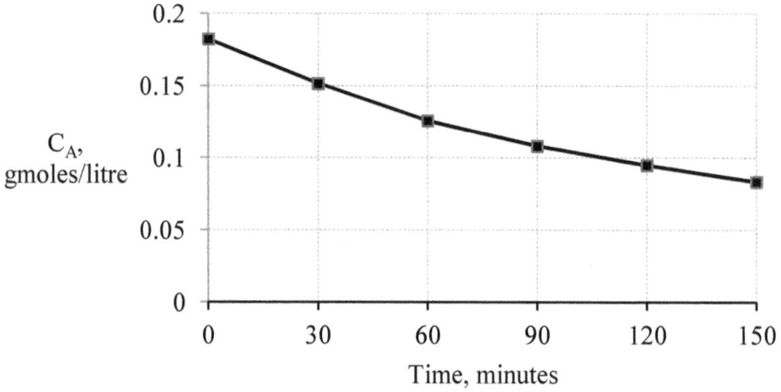

Because the plot is not a straight line, a zero order reaction is eliminated. The slight curvature of the plot, however, does not show clearly whether this is a first or second order reaction. We have to, therefore, try to fit the data into one of these two elementary reaction models.

If the reaction is first order

$$C_A = C_{Ao}e^{-k_1 t} \tag{2}$$

and

$$\ln\left(\frac{C_A}{C_{Ao}}\right) = -k_1 t \tag{3}$$

If $\ln\left(\frac{C_A}{C_{Ao}}\right)$ is plotted against t, the plot should result in a straight line of slope $-k_1$. When we tabulate $\ln\left(\frac{C_A}{C_{Ao}}\right)$ and t, we get the table below

Time, minutes	C_A, mol/litre	C_A/C_{Ao}	ln (C_A/C_{Ao})
0	0.1823	1.000	0
30	0.1515	0.831	-0.185
60	0.1257	0.690	-0.371
90	0.108	0.592	-0.524
120	0.0948	0.520	-0.654
150	0.0833	0.457	-0.783

When we plot $\ln\left(\frac{C_A}{C_{Ao}}\right)$ versus t, as in Fig. 3.21b below, we do not get a straight line.

Fig. 3.21b: Plot of ln (C_A/C_{Ao}) versus t

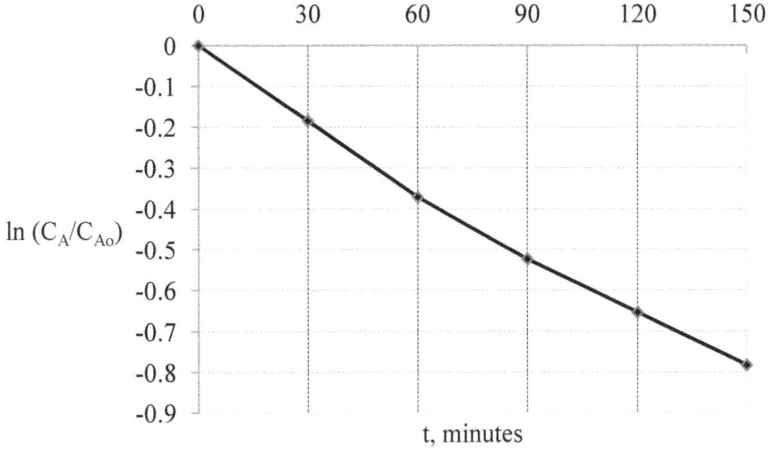

If the reaction is second order

$$\frac{1}{C_A} - \frac{1}{C_{Ao}} = k_2 t \qquad (4)$$

A plot of $\frac{1}{C_A}$versus t should give a straight line. The table below lists $\frac{1}{C_A}$ and t while Fig. 3.21c below it gives the plot of $\frac{1}{C_A}$ versus t. This chart shows that the plot is a straight line with average slope given by

$$k_2 = \frac{12.005 - 5.485}{150 - 0} = 0.0435 \frac{litre}{gmol\,.min}$$

Time, minutes	C_A, mol/litre	$1/C_A$, litre/mol
0	0.1823	5.485
30	0.1515	6.601
60	0.1257	7.955
90	0.108	9.259
120	0.0948	10.549
150	0.0833	12.005

Fig. 3.21c: Plot of $1/C_A$ versus t

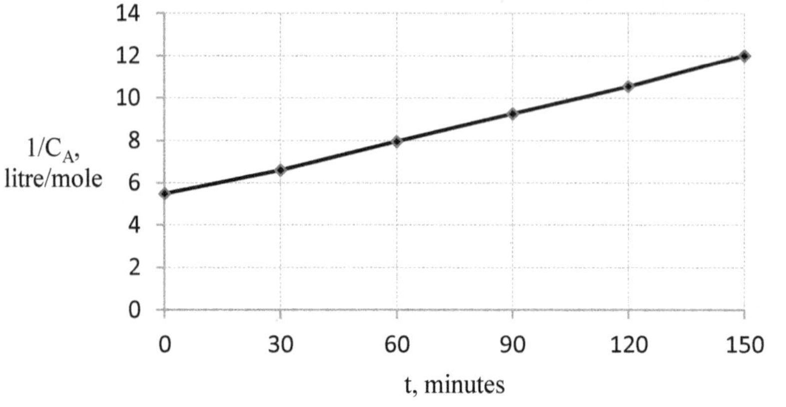

Thus the rate equation is

$$\frac{1}{C_A} - \frac{1}{C_{Ao}} = 0.0435t \qquad \textit{Ans}$$

This may, also, be expressed as

$$0.0435t + 5.485 = \frac{1}{C_A}$$

or as

$$C_A = \frac{1}{0.0435t + 5.485}$$

Example 3.22

When the concentration of A in the simple reaction $A \rightarrow B$ was changed from 0.502 kmol/m^3 to 1.07 kmol/m^3, the half life dropped from 51 seconds to 26 seconds at 26 C. What is the order of the reaction and the value of the rate constant? Take

$$t_{1/2} = \frac{2^{n-1}-1}{k(n-1)} \cdot C_{Ao}{}^{1-n}$$

if you think it applies where

$t_{1/2}$	half life
n	order of the reaction
k	rate constant
C_{Ao}	initial concentration

Answer

Since the half life changed with change in initial concentration, the reaction cannot be first order; that is n ≠ 1. Hence the given equation can be used. Thus, when $C_{Ao} = 0.502$ kmol/m³ and $t_{1/2} = 51$ seconds,

$$51 = \frac{2^{n-1} - 1}{k(n-1)} . (0.502)^{1-n} \qquad (1)$$

When $C_{Ao} = 1.007$ kmol/m³ and $t_{1/2} = 26$ seconds,

$$26 = \frac{2^{n-1} - 1}{k(n-1)} . (1.007)^{1-n} \qquad (2)$$

Dividing (1) by (2)

$$\frac{51}{26} = \left(\frac{0.502}{1.007}\right)^{1-n}$$

$$\log 1.9615 = (1-n)\log(0.4985)$$

from which

$$1 - n = \frac{\log 1.9615}{\log(0.4985)} = \frac{0.2926}{-0.3023} = -0.9679$$

which gives

$$n = 1.9679 \cong 2 \quad Ans.$$

Example 3.23

The rate constant of a first order reaction was determined between 170 C and 220 C as follows

T, C	172.2	187.7	202.6	218.1
Rate constant, 1/s	9.97×10^{-5}	3.01×10^{-4}	7.80×10^{-4}	2.04×10^{-3}

Calculate the activation energy and the pre-exponential factor for the reaction. Take the molar gas constant, R = 8.3143 J/mol.K

Answer

To calculate the activation energy we shall use the Arrhenius theory equation because it is the only one of the two equations that contains a pre-exponential factor.

The equation is

$$k = Ae^{-\Delta E/RT}$$

where

A	pre-exponential factor
E	energy of activation for the reaction
R	ideal gas constant
T	absolute temperature at which the reaction takes place

The calculation can be done either numerically or graphically.

For a graphical solution, we first linearise the Arrhenius equation as follows

$$\log k = \log A - \frac{\Delta E}{2.303RT} \quad (1)$$

A graphical plot of log k versus 1/T should give a straight line of slope $-\frac{\Delta E}{2.303R}$ and intercept $\log A$.

The values used for the plot are calculated as shown in the Table below while the graphical plot is shown inn Fig. 3.23 below.

Table of Values Used in Fig. 3.23

T, C	k, 1/s	T, K	1/T, K^{-1}	Log_{10} k
172.2	9.97×10^{-5}	445.2	2.25×10^{-3}	-4.00
187.7	3.01×10^{-4}	460.7	2.17×10^{-3}	-3.52
202.6	7.80×10^{-4}	475.6	2.10×10^{-3}	-3.11
218.1	2.04×10^{-3}	491.1	2.04×10^{-3}	-2.69

Fig. 3.23: Plot of Log_{10} k versus 1/T

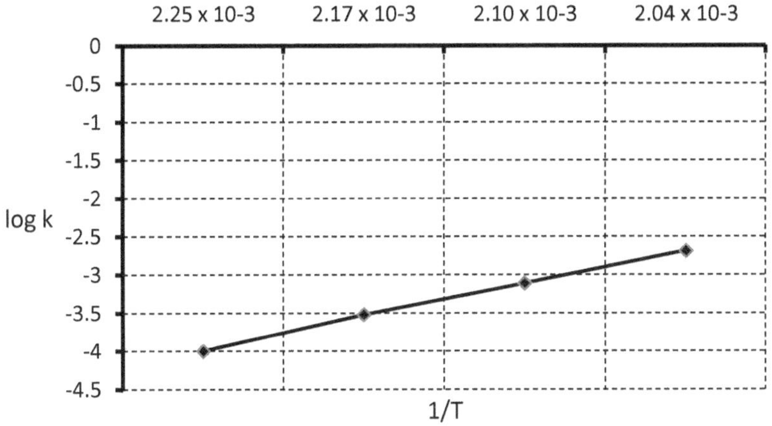

The slope is

$$-\frac{\Delta E}{2.303R} = \frac{-2.69 - (-4.0)}{(2.04 - 2.25)\ x\ 10^{-3}} = -6238.10$$

$$\Delta E = 5762.71\ x\ 2.303\ x\ 8.3143 = 1.19446\ x10^{5}\ J/mol \qquad (2)$$

At 172.2 C, from equation (1), (2) and the values in the Table

$$-4.0 = \log A - \frac{1.19446\ x\ 10^{5}}{2.303\ x\ 8.3143\ x\ 445.2}$$

That is

$$\log A = 8.9440$$

and

$$A = 1.028\ x\ 10^{10}, s^{-1} \qquad (3)$$

In industrial practice, the raw data are plotted directly in a log-linear graph paper. These papers are, however, not much in general everyday use anymore.

The values of A and ΔE can also be calculated by the solution of simultaneous equations or by least squares analysis.

The problem with using the method of simultaneous equations, is that, since there are two unknowns in the Arrhenius equation, we can use, at any one time, only two of the four sets of data given.

Each set of two will give different values of A and ΔE because of the variability associated with experimental data. For example, if we use the first and last values at 172.2 C and 218.1 C in equation (1).

$$-4 = \log A - 1.173\,x\,10^{-4}\Delta E \tag{4}$$

$$-2.69 = \log A - 1.063\,x\,10^{-4}\Delta E \tag{5}$$

Subtracting equation (4) from equation (5)

$$1.31 = 1.1\,x\,10^{-5}\Delta E \quad \text{or} \quad \Delta E = 1.1909\,x\,10^{5}\,J/mol$$

Substituting this value in (4)

$$\log A = -4 + 1.173\,x\,10^{-4}\,x\,1.1909\,x\,10^{5} = 9.9693$$

giving

$$A = 9.3166\,x\,10^{9}\,s^{-1}$$

This value of ΔE differs by +8 % and that of A by 9.4 % from the values obtained by graphical procedure. The level of error depends on which combination of sets of data is used. Values closer to those obtained graphically will be obtained if the simultaneous equations are solved for every combination of pairs of data and the average of all the results taken.

The mathematical procedure that eliminates this tedium and gives a smoothed average result is the least squares regression method attributed to Gauss (see the Appendix for details).

If equation (1) is rewritten as

$$\log k = \log A - \frac{\Delta E}{2.303R}\cdot\frac{1}{T} = y = a + bx \tag{6}$$

where $y = \log k$, $a = \log A$, $b = -\frac{\Delta E}{2.303R}$, and $x = \frac{1}{T}$, the least squares regression gives

$$a = \frac{\sum_1^4 y \sum_1^4 x^2 - \sum_1^4 x \sum_1^4 xy}{4\sum_1^4 x^2 - (\sum_1^4 x)^2} \tag{7}$$

$$b = \frac{4\sum_1^4 xy - \sum_1^4 x \sum_1^4 y}{4\sum_1^4 x^2 - (\sum_1^4 x)^2} \tag{8}$$

A Table is constructed according to equations (7) and (8)

T, K	**x = 1/T, K^{-1}**	**y = log$_{10}$ k**	**xy**	**x^2**
445.2	2.25 x 10^{-3}	-4.00	-0.009	5.0625 x10^{-6}
460.7	2.17 x 10^{-3}	-3.52	-0.0076384	4.7089 x 10^{-6}
475.6	2.10 x 10^{-3}	-3.11	-0.006531	4.41 x 10^{-6}
491.1	2.04 x 10^{-3}	-2.69	-0.0054876	4.1616 x 10^{-6}
Totals	8.56 x 10^{-3}	-13.32	-0.028657	1.8343 x 10^{-5}

$$a = \frac{\sum_1^4 y \sum_1^4 x^2 - \sum_1^4 x \sum_1^4 xy}{4\sum_1^4 x^2 - (\sum_1^4 x)^2}$$

$$= \frac{-13.32\ x\ 1.8343\ x\ 10^5 + 8.56\ x\ 10^{-3}\ x\ 0.028657}{4\ x\ 1.8343\ x\ 10^5 - (8.56\ x\ 10^{-3})^2}$$

$$= 9.9102 \tag{9}$$

$$b = \frac{4\sum_1^4 xy - \sum_1^4 x \sum_1^4 y}{4\sum_1^4 x^2 - (\sum_1^4 x)^2} = \frac{-4\ x\ 0.028657 + 0.00856\ x\ 13.32}{4\ x\ 1.8343\ x\ 10^5 - (8.56\ x\ 10^{-3})^2}$$

$$= -6186.99 \tag{10}$$

From (9) and (10)

$$a = \log A = 9.9102 \quad or \quad A = 8.132\ x\ 10^9$$

$$b = -\frac{\Delta E}{2.303R} = -6186.99 \quad or \quad \Delta E = 118467.45\ J/mol$$

These results are summarised below. They do not differ much from each other except by the tedium of processing. Which method is used depends on the quality of the data and other considerations of time and economy.

	Graphical Procedure	Simultaneous Equation	Linear Regression
A, per second	1.028×10^{10}	9.317×10^{9}	8.132×10^{9}
ΔE, J/mol	1.195×10^{5}	1.191×10^{5}	1.185×10^{5}

Example 3.24

The Arrhenius equations for the rate of decomposition of methyl nitrite and ethyl nitrite are

$$k_1, s^{-1} = 10^{13} exp\left(-\frac{152300}{RT}\right)$$

and

$$k_2, s^{-1} = 10^{14} exp\left(-\frac{157700}{RT}\right)$$

respectively. At what temperature will the two rate constants be equal?

Answer

If the two rate constants are equal

$$k_1 = 10^{13} exp\left(-\frac{152300}{RT}\right) = k_2 = 10^{14} exp\left(-\frac{157700}{RT}\right)$$

That is

$$exp\left(-\frac{152300}{RT} + \frac{157700}{RT}\right) = 10$$

$$exp\left(\frac{5400}{RT}\right) = 10 \quad \text{or} \quad \frac{5400}{RT} = 2.3026$$

If R = 8.3143 J/mol.K, then

$$T = \frac{5400}{8.3143 \ x \ 2.3026}, \frac{J}{mol}\frac{mol.K}{J} = 282.07\ K \quad Ans$$

Example 3.25

The first order reversible, liquid, reaction, $A \leftrightarrow R$, with $C_{Ao} = 0.5$ mol/litre, $C_{Ro} = 0$, takes place in a batch reactor. After 8 minutes, conversion of A is 33.3 % while equilibrium conversion is 66.7 %. Find the rate equation for this reaction.

Answer

For a constant volume batch reaction $A \rightleftharpoons R$

$$r = -\frac{1}{V}\frac{dn_A}{dt} = -\frac{dC_A}{dt} = k_1 C_A - k_2 C_R \tag{1}$$

where k_1 is the rate constant for the forward reaction and k_2 is the rate constant for the backward reaction.

This is equivalent to

$$-\frac{dC_{Ao}(1 - X_A)}{dt} = k_1 C_{Ao}(1 - X_A) - k_2 C_{Ao}(\theta_R + X_A) \tag{2}$$

where

$$\theta_R = \frac{C_{Ro}}{C_{Ao}} \tag{3}$$

Thus

$$\frac{dX_A}{dt} = k_1 - k_2\theta_R - X_A(k_1 + k_2) \tag{4}$$

If the equilibrium constant, based on concentration, K_C, is

$$K_C = \frac{k_1}{k_2} \tag{5}$$

At equilibrium, the equilibrium conversion, X_{Ae}, is obtained from

$$\frac{dX_A}{dt} = k_1 - k_2\theta_R - X_{Ae}(k_1 + k_2) = 0 \tag{6}$$

and

$$X_{Ae} = \frac{K_C - \theta_R}{K_C + 1} \tag{7}$$

If we subtract equation (6) from equation (4), we can get a general expression which relates any fractional conversion to the equilibrium conversion

$$\frac{dX_A}{dt} = (k_1 + k_2)(X_{Ae} - X_A) \tag{8}$$

When $C_{Ro} = 0$, equation (3) gives that $\theta_R = 0$. For $X_{Ae} = 0.667$, equation (7) gives

$$0.667 = \frac{K_C - 0}{K_C + 1}$$

from which

$$K_C = 2.003 \tag{9}$$

From equation (8)

$$\int_0^{X_A} \frac{dX_A}{0.667 - X_A} = (k_1 + k_2)t$$

That is

$$\ln \frac{0.667}{0.667 - X_A} = (k_1 + k_2)t$$

At 8 minutes and 33.3 % conversion

$$\ln \frac{0.667}{0.667 - 0.333} = (k_1 + k_2)\, x\, 8$$

which gives

$$k_1 + k_2 = 0.08646 \tag{10}$$

From (9)

$$K_C = 2.003 = \frac{k_1}{k_2} \quad \text{or} \quad k_1 = 2.003k_2 \tag{11}$$

Substituting (11) in (10)

$$3.003k_2 = 0.08646 \quad \text{or} \quad k_2 = 0.02879 \tag{12}$$

and

$$k_1 = 2.003\, x\, 0.02879 = 0.05767 \tag{13}$$

The rate equation is, therefore, from (8), (12) and (13)

$$\frac{dX_A}{dt} = (0.05767 + 0.02879)(0.667 - X_A)$$

$$\frac{dX_A}{dt} = 0.05767 - 0.08646X_A \ Ans$$

Example 3.26

The gas phase reaction $A = B + 2C$ was carried out in a constant volume batch reactor. Runs 1 to 7 were carried out at 100 C while run 8 was carried out at 110 C. Determine, from the data below, the reaction order, the reaction rate constant and the activation energy for this reaction.

Run	Initial concentration, C_{Ao}, gmol/litre	Half life, minutes
1	0.020	5.64
2	0.030	3.67
3	0.010	9.80
4	0.050	1.96
5	0.070	1.43
6	0.040	2.82
7	0.060	1.70
8	0.025	2.00

Answer

The half life of a zero order reaction is given by

$$\tau_{½} = \frac{C_{Ao}}{2k_o} \tag{1}$$

for a first order reaction by

$$\tau_{½} = \frac{\ln 2}{k_1} \tag{2}$$

and for a second order reaction by

$$\tau_{½} = \frac{1}{k_2 C_{Ao}} \tag{3}$$

These equations show that the half life is directly proportional to the initial concentration for a zero order reaction, is independent of initial concentration for a first order reaction and is inversely proportional to the initial concentration for a second order reaction.

Since the given data shows that the half life is dependent on initial concentration, a first order reaction is ruled out. We are left, therefore, with either a zero order or second order reaction. A graphical plot of the given data on half life versus initial concentration will enable us determine whether the reaction is zero or second order. If it is a straight line with slope greater or less than zero, it will indicate that the reaction is zero order. If it has curvature, it would indicate that the reaction is second order

Fig. 3.26: Plot of Half Life versus Initial Concentration

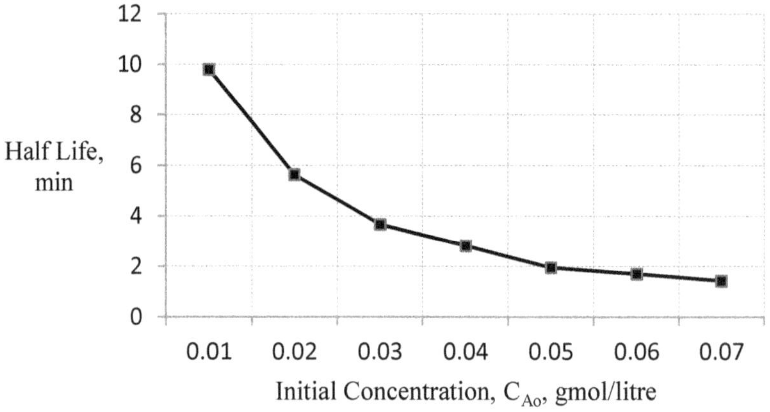

. As the graphical plot of Fig. 3.26 shows, the reaction is second order.

Since equation (3) shows that the product of the half life and C_{Ao} will be constant equal to $1/k_2$ at 100 C, we can save the labour of plotting another graph and construct the table below

Initial concentration, C_{Ao}, gmol/litre	**Half life, minutes**	**$\tau_{½}C_{Ao}$, gmol/litre.min**
0.010	9.8	0.098
0.020	5.64	0.1128
0.030	3.67	0.1101
0.040	2.82	0.1128
0.050	1.96	0.098
0.060	1.70	0.1020
0.070	1.43	0.1001

This gives us an average of

$$\frac{1}{k_2} = \frac{0.098 + 0.1128 + 0.1101 + 0.1128 + 0.098 + 0.1020 + 0.1001}{7}$$
$$= 0.1048$$

That is, at 100 C

$$k_2 = \frac{1}{0.1048} = 9.54 \ \frac{litre.min}{gmol} \tag{4}$$

At 110 C, from the data given and equation (3)

$$k_2 = \frac{1}{\tau_{½}C_{Ao}} = \frac{1}{2\,x\,0.025} = 20 \ \frac{litre.min}{gmol} \tag{5}$$

Since

$$k = Ae^{-\Delta E/RT}$$

At 100 C = 373 K, $$k_{100} = 9.54 = Ae^{-\Delta E/373R} \tag{6}$$

At 110 C = 383 K, $$k_{110} = 20.00 = Ae^{-\Delta E/383R} \tag{7}$$

Dividing equation (7) by equation (6)

$$\frac{k_{110}}{k_{100}} = \frac{20}{9.54} = \frac{e^{-\Delta E/383R}}{e^{-\Delta E/373R}} = e^{-\frac{\Delta E}{R}\left(\frac{1}{383}-\frac{1}{373}\right)} = e^{7.0\,x\,10^{-5}\frac{\Delta E}{R}}$$

That is

$$e^{7.0\,x\,10^{-5}\frac{\Delta E}{R}} = 2.0964$$

$$7.0\,x\,10^{-5}\frac{\Delta E}{R} = 0.7402$$

Since R = 8.314 J/mol.K

$$\Delta E = \frac{0.7402\,x\,8.314}{7.0x\,10^{-5}} = 87{,}914.6\frac{J}{mol} \quad Ans.$$

References for Chapter Three

1. B. N. Nnolim: Unpublished Lecture Notes in Chemical Reaction Engineering; IMT, Enugu, Nigeria; 1989
2. *Catalyst Bulletin, 1980*; Harshaw Chemical Company, Ohio, USA
3. *Catalyst Bulletin, 1980*; Harshaw Chemical Company, Ohio, USA
4. Denbigh K. G. & Turner J. C. R.; *Chemical Reactor Theory*; 2nd edition, London; Cambridge University Press, 1971
5. Fogler, H Scott, *The Elements of Chemical Kinetics and Reactor Calculations*; Prentice Hall Inc; N. J., USA, 1974
6. Levenspiel O; *Chemical Reaction Engineering*; Wiley International Edition; New York, USA, 1972
7. Walas S. M.; *Reaction Kinetics for Chemical Engineers*; McGraw-Hill Book Company, New York, USA, 1959

CHAPTER FOUR: MATERIAL BALANCES IN CHEMICAL REACTORS

Example 4.1

State the equation of the general material balance around a chemical reactor, indicating the meaning of the terms involved.

Using this equation, develop the concepts and major characteristics of three ideal reactors.

If a given reaction is to be carried out, what would be the considerations in choosing a suitable reactor?

Answer

The general material balance around a chemical reactor may be derived by considering the input, output, generation/consumption and accumulation associated with any chemical reactor as shown in the sketch below

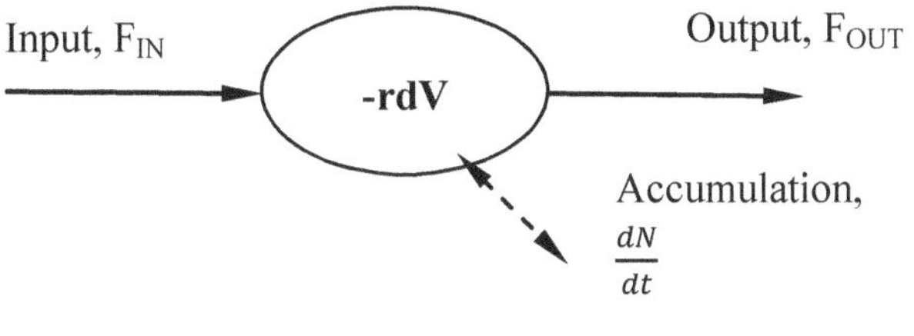

By the law of conservation of matter

$$Input + Generation/consumption = Output + Accumulation \quad (1)$$

In a chemical reactor with a reaction volume, V, in which material inflow per unit time is F_{IN}, material outflow per unit time, F_{OUT}, and in which reaction is occurring at the rate of r moles per unit time per unit volume of reaction mixture, equation (1) may be expressed in symbolic mathematical form as

$$F_{IN} + \int_0^V rdV = F_{OUT} + \frac{dN}{dt} \quad (2)$$

where N is the number of moles of the reactant of interest at time t. This is the mathematical expression of the general material balance around a chemical reactor.

a. For a batch reactor, in which $F_{IN} = 0$; $F_{OUT} = 0$, equation (2) becomes

$$\frac{dN}{dt} = \int_0^V rdV \quad (3)$$

If this batch reactor is perfectly mixed so that *r* is not a function of position in the reactor

$$\frac{dN}{dt} = rV \quad or \quad \frac{1}{V}\frac{dN}{dt} = \frac{dC}{dt} = r \quad (4)$$

This is the reaction rate expression for a constant volume batch reactor.

b. For a plug flow reactor, if flow is assumed to be steady state so that there is no accumulation; that is $\frac{dN}{dt} = 0$. Equation (2) now becomes

$$\int_0^V rdV = F_{OUT} - F_{IN} \quad (5)$$

But

$$dF = F_{OUT} - F_{IN} = rdV$$

so that

$$V = \int \frac{dF}{r} \quad (6)$$

This is the reaction rate expression for a plug flow reactor.

c. For a continuous stirred tank reactor (CSTR), if there is steady state flow, $\frac{dN}{dt} = 0$. Additionally, if the reactor is perfectly mixed so that *r* is not a function of position in the reactor $\int_0^V rdV = rV$. Equation (2), then, becomes

$$F_{IN} + rV = F_{OUT}$$

from which

$$V = \frac{F_{OUT} - F_{IN}}{r} \qquad (7)$$

This is the reaction rate expression for a continuous stirred tank (CSTR) reactor.

The consideration for choosing a reactor will thus depend on whether the material processing conditions or requirements satisfy the restrictions which describe a batch, CSTR or PFR.

Example 4.2

100 moles of NO_2 is to be converted to N_2O_5 by contacting with 100 moles of air according to the reaction

$$2NO_2 + O_2 = N_2O_5$$

Develop the stoichiometric table for this reaction given that the conversion is 15 %. Assume that air contains 21 % oxygen and 79 % nitrogen. Nitrogen does not take part in the reaction.

If the volume of the reaction mixture, at any time, is V, given by

$$V = V_o(1 - 0.275X)$$

Determine the concentration of all the species at 15 % conversion.

Answer

For the given reaction, take NO_2 as the key reactant and the fractional conversion, X, based on it. The stoichiometric table for the reaction is then as follows;

Basis: 100 moles of initial reaction Mixture, X = 0.15

Species	Moles initially present	Moles reacted	Moles remaining
NO_2	100	-0.15 x 100 = -15	85
O_2	21	-21 x 0.15 = -3.75	17.25
N_2O_5	0	0.15 x 100 x 0.5 = 7.5	7.50
N_2	79	-	79
Totals	200		188.75

The concentrations are, therefore

For NO_2

$$C_{NO_2} = \frac{85}{V_o(1 - 0.275\,x\,0.15)} = \frac{88.66}{V_o} \qquad Ans$$

For O_2

$$C_{O_2} = \frac{17.25}{V_o(1 - 0.275\,x\,0.15)} = \frac{17.99}{V_o} \qquad Ans$$

For N_2O_5

$$C_{N_2O_5} = \frac{7.5}{V_o(1 - 0.275\,x\,0.15)} = \frac{7.82}{V_o} \qquad Ans$$

For N_2

$$C_{N_2} = \frac{79}{V_o(1 - 0.275\,x\,0.15)} = \frac{82.40}{V_o} \qquad Ans$$

Example 4.3

An initial reaction mixture of 60 % acetaldehyde and 40 % inerts is converted, at constant temperature and pressure, to methane and carbon monoxide according to the reaction

$$CH_3CHO \rightarrow CH_4 + CO$$

Determine the percent volume change for 90 % conversion of acetaldehyde.

Answer

The volume change is given by

$$V = V_o(1 + \epsilon X) \qquad (1)$$

where V_o is the initial volume of the reaction mixture, X is the fractional conversion and **ϵ** is given by

$$\epsilon = \delta y_{Ao}$$

δ is the net change in the stoichiometric coefficient, and y_o is the initial mole fraction.

For the given reaction

$$\delta = \frac{1+1-1}{1} = 1$$

Initial mole fraction, y_{Ao}, is 0.6. Hence, **ϵ** is given by

$$\epsilon = \delta y_o = 1 \; x \; 0.6 = 0.6$$

When these are substituted into equation (1)

$$V = V_o(1 + 0.6X)$$

At 90 % conversion,

$$V = V_o(1 + 0.6 \; x \; 0.9) = 1.54V_o$$

Percent volume change is therefore, 54 % by inspection or mathematically

$$Percent\ volume\ change = \frac{V - V_o}{V_o} \; x \; 100 = \frac{1.54V_o - V_o}{V_o} \; x \; 100$$

$$= 54\ \% \qquad Ans$$

Example 4.4

The saponification reaction, at constant volume, for the formation of soap from aqueous caustic soda and glycerol stearate is

$$3NaOH + (C_{17}H_{35}COO)_3C_3H_5 = 3C_{17}H_{35}COONa + C_3H_5(OH)_3$$
$$\qquad\qquad\qquad\qquad soap \qquad\qquad glycerine$$

Set up a stoichiometric table expressing the concentration of each species in terms of its initial concentration and its subsequent conversion.

If the initial mixture consisted of 10 gmol per litre of sodium hydroxide and 2 gmol per litre of of glycerol stearate, what will the concentration of glycerine be when the conversion of sodium hydroxide is 20 %?

Answer

Because of the complicated nature of the formulae of the compounds involved, let us represent caustic soda simply as A, glycerol stearate as B, soap as C and glycerine as D. Then, the given equation becomes

$$3A + B = 3C + D$$

That is

$$A + ⅓B = C + ⅓D \qquad (1)$$

The stoichiometric table is then built up from equation (1) as shown below

Species	Gmoles initially present	Gmoles reacted	Gmoles remaining
A	N_{Ao}	$-N_{Ao}X_A$	$N_{Ao}(1 - X_A)$
B	N_{Bo}	$-⅓N_{Ao}X_A$	$N_{Bo} - ⅓N_{Ao}X_A$
C	N_{Co}	$N_{Ao}X_A$	$N_{Co} + N_{Ao}X_A$
D	N_{Do}	$⅓N_{Ao}X_A$	$N_{Do} + ⅓N_{Ao}X_A$
Inerts	N_{Io}	-	N_{Io}
Total	N_{To}		$N_{To} + (⅓ + 1 - ⅓ - 1)N_{Ao}X_A = N_{To}$

Since

$$\delta = ⅓ + 1 - ⅓ - 1 = 0$$

$$\epsilon = \delta y_{Ao} = 0 \qquad (2)$$

Hence

$$C_A = \frac{C_{Ao}(1 - X_A)}{(1 + \epsilon X_A)} = C_{Ao}(1 - X_A) \qquad Ans$$

$$C_B = \frac{C_{Bo} - ⅓C_{Ao}X_A}{(1 + \epsilon X_A)} = C_{Bo} - ⅓C_{Ao}X_A \qquad Ans$$

$$C_C = \frac{C_{Co} + C_{Ao}X_A}{(1 + \epsilon X_A)} = C_{Co} + C_{Ao}X_A \qquad Ans$$

$$C_D = \frac{C_{Do} + ⅓C_{Ao}X_A}{(1 + \epsilon X_A)} = C_{Do} + ⅓C_{Ao}X_A \qquad Ans$$

If $C_{Ao} = 10\ gmol/l$, $C_{Bo} = 2\ gmol/l$, $C_{Co} = 0\ gmol/l$,

$C_{Do} = 0\ gmol/l$, and $X_A = 0.2$, then

$$C_D = C_{Do} + ⅓C_{Ao}X_A = 0 + ⅓\ x\ 10\ x\ 0.2 = 0.67\frac{gmol}{l} \qquad Ans$$

Example 4.5

Cumene decomposes over a solid catalyst to form benzene and propylene according to the reaction

$$C_6H_5CH(CH_3)_2 \rightarrow C_6H_6 + C_3H_6$$

For an equal molar feed of cumene and nitrogen, set up a stoichiometric table and then express the concentration of each species in terms of conversion and the initial concentration of cumene when the reaction is carried out at constant pressure in a constant volume reactor

Answer

Let X_C be the fractional conversion of cumene.

In both a constant volume reactor or at constant pressure, the stoichiometric table, in terms of kmoles and fractional conversion, is given as follows

Basis: 100 kmol Cumene

Species	**kmoles initially present**	**kmoles reacted or formed**	**kmoles remaining**
Cumene	100	$-100X_C$	$100(1-X_C)$
Benzene	0	$100X_C$	$100X_C$
Propylene	0	$100X_C$	$100X_C$
Nitrogen	100	-	100
Total	200		$200+100X_C$

For a real gas, at any time t, the reaction conditions are defined by $Z_t,\ P_t,\ V_t,\ T_t$. That is

$$P_t\, V_t\ = Z_t N_t R\, T_t \tag{1}$$

At the initial conditions they are defined by Z_o, P_o, V_o, T_o so that

$$P_o\, V_o\ = Z_o N_o R T_o \tag{2}$$

from which

$$V_t = V_o . \frac{Z_t}{Z_o} . \frac{N_t}{N_o} . \frac{T_t}{T_o} . \frac{P_o}{P_t} \tag{3}$$

As can be seen from the stoichiometric table

$$N_t = N_o + \delta N_{io} X_i \tag{4}$$

where **δ** is the increase in the total number of moles of the gas phase per mole of key reactant reacted, N_o is the total number of moles of reactants, products and inerts, initially, present and N_{io} is the initial number of moles of reactant i. Then, from (3) and (4)

$$V_t = V_o \frac{N_o+\delta N_{io} X_i}{N_o} \frac{Z_t}{Z_o} \frac{T_t}{T_o} \frac{P_o}{P_t} = V_o(1+\delta . y_{Co} X_i) \frac{Z_t}{Z_o} . \frac{T_t}{T_o} . \frac{P_o}{P_t} \tag{5}$$

where $y_{Co} = \frac{N_{io}}{N_o}$, the initial mole fraction of component i

If we represent ϵ as

$$\epsilon = \delta . y_{Co} \tag{6}$$

then, the reaction volume, V_t, will be given by

$$V_t = V_o(1 + \epsilon X_i)\frac{Z_t}{Z_o}\frac{T_t}{T_o}\frac{P_o}{P_t} \tag{7}$$

For an ideal gas at constant pressure, $\frac{Z_t}{Z_o} = 1$, $\frac{P_o}{P_t} = 1$

$$V_t = V_o(1 + \epsilon X_i)\frac{T_t}{T_o} \tag{8}$$

For the reaction given

$$\delta = 1 + 1 - 1 = 1 \tag{9}$$

$$y_{Co} = \frac{100}{200} = 0.5 \tag{10}$$

so that

$$\epsilon = \delta . y_{Co} = 0.5 \tag{11}$$

Thus, from (8) and (11)

$$V_t = V_o(1 + 0.5X_i)\frac{T_t}{T_o} \tag{12}$$

When this is applied to the stoichiometric table, we obtain the concentrations as shown below

Species	**Concentration at time t**	**Concentration at time t**
Cumene	$\frac{100(1 - X_C)}{V_o(1 + 0.5X_C)}\frac{T_o}{T}$	$\frac{C_{Co}(1 - X_C)T_o}{(1 + 0.5X_C)T}$
Benzene	$\frac{100X_C}{V_o(1 + 0.5X_C)}\frac{T_o}{T}$	$\frac{C_{Co}X_C T_o}{(1 + 0.5X_C)T}$
Propylene	$\frac{100X_C}{V_o(1 + 0.5X_C)}\frac{T_o}{T}$	$\frac{C_{Co}X_C T_o}{(1 + 0.5X_C)T}$
Nitrogen	$\frac{100}{V_o(1 + 0.5X_C)}\frac{T_o}{T}$	$\frac{C_{Co}T_o}{(1 + 0.5X_C)T}$
Total	$\frac{(200 + 100X_C)}{V_o(1 + 0.5X_C)}\frac{T_o}{T}$	$\frac{C_{Co}(2 + X_C)T_o}{(1 + 0.5X_C)T}$

Example 4.6

The following equation represents the reaction, in a batch reactor, between compounds A and B, each in a water solution.

$$3A + B = 3C + D$$

Set up a stoichiometric table describing the concentration of each species in terms of its initial concentration and its fractional conversion. Assume that water is the inert compound.

If the initial concentration of A and B are 10 kmol/m^3 and 2 kmol/m^3, respectively, determine the concentration of D when the conversion of A is 20 %.

Answer

For clarity in the stoichiometric table, it is usual to express it in terms of a definite unit of the key reactant such as A. We can, if we regard A as the key reactant, thus, express the given reaction equation as

$$A + \frac{1}{3}B = C + \frac{1}{3}D \qquad (1)$$

We can, also, construct a stoichiometric table, based on equation (1) as follows

Specie	**Moles initially present**	**Change due to reaction**	**Moles remaining**	**Concentration, kmol/m^3**
A	N_{Ao}	$-N_{Ao}X_A$	$N_{Ao} - N_{Ao}X_A$	$C_{Ao}(1 - X_A)$
B	N_{Bo}	$-\frac{1}{3}N_{Ao}X_A$	$N_{Bo} - \frac{1}{3}N_{Ao}X_A$	$C_{Ao}\left(\theta_B - \frac{X_A}{3}\right)$
C	N_{Co}	$N_{Ao}X_A$	$N_{Co} + N_{Ao}X_A$	$C_{Ao}(\theta_C + X_A)$
D	N_{Do}	$\frac{1}{3}N_{Ao}X_A$	$N_{Do} + \frac{1}{3}N_{Ao}X_A$	$C_{Ao}\left(\theta_D + \frac{X_A}{3}\right)$
Inerts	N_{Io}	0	N_{Io}	C_{Io}

where

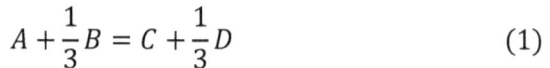

$$C_{Ao} = \frac{N_{Ao}}{V_o}; \ \theta_B = \frac{C_{Bo}}{C_{Ao}}; \ \theta_C = \frac{C_{Co}}{C_{Ao}}; \ \theta_D = \frac{C_{Do}}{C_{Ao}};$$

$$V_o = \ constant \ initial \ volume \ of \ reaction \ mixture \qquad (3)$$

For $C_{Ao} = 10\frac{kmol}{m^3}$; $C_{Bo} = 2\frac{kmol}{m^3}$; $X_A = 0.2$; and $\theta_C = 0 = \theta_D$;

we get from the stoichiometric table that

$$C_D = C_{Ao}\left(\theta_D + \frac{X_A}{3}\right) = C_{Ao}\left(\frac{X_A}{3}\right) = 10\,x\,\frac{0.2}{3} = 0.66\,\frac{kmol}{m^3} \qquad Ans$$

Example 4.7

The liquid reaction A→R was carried out in a batch reactor and the following data obtained.

C_A, mol/litre	$-r_A$, mol/litre.min
0.1	0.1
0.2	0.3
0.4	0.6
0.6	0.25
0.8	0.06
1.0	0.05
1.2	0.047
1.4	0.045
1.6	0.044
1.8	0.043
2	0.042

How long must the reaction be continued for the concentration to drop from C_{Ao} = 1.4 mol/litre to C_A = 0.2 mol/litre?

Answer

For a batch reactor

$$t = C_{Ao}\int_{X_{A1}}^{X_{A2}} \frac{dX_A}{-r_A} = \int_{C_{A1}}^{C_{A2}} \frac{dC_A}{-r_A} \tag{1}$$

where

$$X_A = \frac{C_{Ao} - C_A}{C_{Ao}} \tag{2}$$

To integrate equation (1) we need to determine the limits of integration C_{A2} and C_{A1}. This we have been given as C_{A1} = 1.4 mol/litre and C_{A2} = 0.2 mol/litre.

If we plot $1/-r_A$ versus C_A, the area under the curve between C_{A1} and C_{A2} will give us t. Thus, we construct a table of $1/-r_A$ versus C_A as shown below.

C_A, mol/litre	$-r_A$, mol/litre.min	$1/-r_A$, litre.min/mol
0.1	0.1	10.00
0.2	0.3	3.33
0.4	0.6	1.67
0.6	0.25	4.00
0.8	0.06	16.67
1.0	0.05	20.00
1.2	0.047	21.28
1.4	0.045	22.22
1.6	0.044	22.73
1.8	0.043	23.26
2	0.042	23.81

When this is plotted, the figure below results.

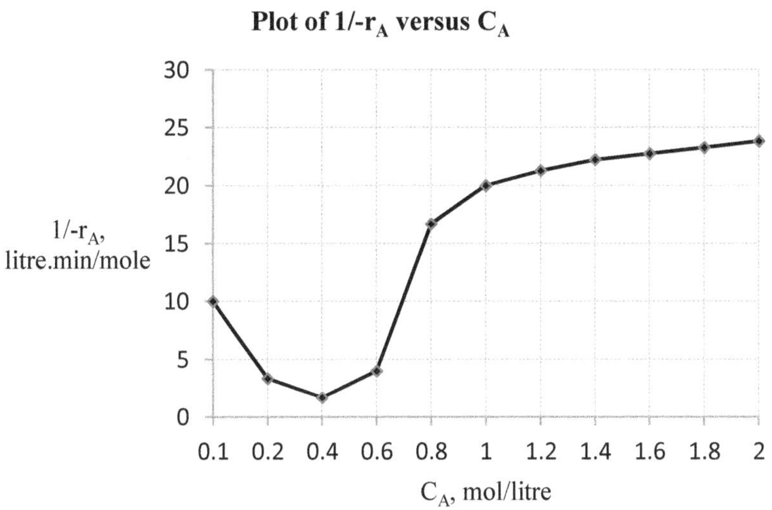

To find the area under the curve between the limits $C_A = 0.2$ to $C_A = 1.4$, let us use Simpson's rule which states that the area under a curve is given by

$$Area = \frac{h}{3}(y_o + 4y_1 + y_2) \tag{3}$$

where

$$h = \frac{y_2 - y_o}{2} \tag{4}$$

The larger the h, the less accurate the integration is so that it is usual to break up the x-axis intervals into many smaller pairs of elements each of which is summed.

If we choose three pairs of elements, we can tabulate them and their Simpson's rule integration as follows

	C_A, mol/litre	$1/-r_A$, litre.min/mol	Area by Simpson's rule, minutes
First pair of elements	0.2	3.33	
	0.4	1.67	
	0.6	4.00	$\frac{0.2}{3}(3.33 + 4\,x 1.67 + 4) = 0.934\,min$

Second pair of elements	0.6	4.00	
	0.8	16.67	
	1.0	20.00	$\frac{0.2}{3}(4 + 4\,x\,16.67 + 20)$ $= 6.045\,min$
Third pair of elements	1.0	20.00	
	1.2	21.28	
	1.4	22.22	$\frac{0.2}{3}(20 + 4\,x\,21.28 + 22.22)$ $= 8.489\,min$
			$Total = 0.934 + 6.045 + 8.489 = 15.468\,minutes$

Thus, the time required is 15.47 minutes. Ans

Example 4.8

1000 mol/h of a substance A at an initial concentration, $C_{Ao} = 1.2$ mol/litre, enters a CSTR and leaves at a concentration, $C_{Af} = 0.3$ mol/litre. If the conversion is 75 %, determine the volume of the CSTR if the rate of reaction, $-r_A = 0.5$ mol/litre.min.

Answer

For a CSTR,

$$V = \frac{F_{Ao}X_A}{-r_A}$$

where F_{Ao} is the entering flow rate, X_A is the fractional conversion of the exit stream and $-r_A$ is the reaction rate within the reactor.

Substituting the given values

$$V = \frac{F_{Ao}X_A}{-r_A} = \frac{1000\,x\,0.75}{0.5\,x\,60},\frac{mol}{h}.\frac{litre.min}{mol}.\frac{h}{min} = 25\,litres.\;Ans$$

Example 4.9

At 600 K, the elementary gas phase reaction

$$C_2H_4 + Br_2 \rightleftharpoons C_2H_4Br_2$$

takes place in a plug flow reactor. 600 m^3/h of gas containing 60 % Br_2, 30 % C_2H_4 and 10 % inerts by volume, are fed into the reactor at 600 K and 1.5 atm. Determine the concentrations of C_2H_4, Br_2 and $C_2H_4Br_2$ at any conversion X of C_2H_4.

Assume that 1 kmol of any ideal gas at STP occupies 22.41 m^3.

Answer

For the given reaction, take C_2H_4 as component A and the key reactant, Br_2 as B, $C_2H_4Br_2$ as C and inerts as I. If the fractional conversion of the key reactant is designated as X_A, the stoichiometric table for the reaction is then as follows;

Basis: 100 moles of initial reaction Mixture

Species	Moles initially present	Moles reacted	Moles remaining
A	30	$-30X_A$	$30(1-X_A)$
B	60	$-30X_A$	$30(2-X_A)$
C	0	$30X_A$	$30X_A$
I	10	-	10
Totals	100		$100-30X_A$

Net change in the stoichiometric coefficient, **δ**, is

$$\delta = \frac{1-1-1}{1} = -1$$

Initial mole fraction, y_{Ao}, is 0.3. Hence, ϵ is given by

$$\epsilon = \delta y_{Ao} = -1 \, x \, 0.3 = \; -0.3$$

The volume of the reaction mixture is, therefore,

$$V = V_o(1 - 0.3X_A) = 600(1 - 0.3X_A)$$

For an ideal gas, $PV = nRT$ so that at any temperature and pressure, T_1 and P_1

$$N_1RT_1 = P_1V_1$$

At STP, with temperature and pressure T_o and P_o,

$$N_oRT_o = P_oV_o$$

Thus

$$\frac{N_1RT_1}{N_oRT_o} = \frac{P_1V_1}{P_oV_o}$$

From which we get that

$$N_1 = \frac{P_1V_1}{P_oV_o} \cdot \frac{T_o}{T_1} N_o = \frac{1.5 \, x \, 600}{1 \, x \, 22.4} \cdot \frac{273}{600} \, x \, 1 = 18.28 \, gmol/h$$

We can now construct the concentration table for 18.28 gmol/h as shown below

Species	Initial concentration, gmol/litre	Concentration at Conversion X_A, gmol/litre
A	$\frac{0.3 \, x \, 18.28}{600} = 0.00914$	$\frac{0.3 \, x \, 18.28(1 - X_A)}{600(1 - 0.3X_A)} = \frac{9.14 \, x \, 10^{-3}(1 - X_A)}{(1 - 0.3X_A)}$
B	$\frac{0.6 \, x \, 18.28}{600} = 0.01828$	$\frac{9.14 \, x \, 10^{-3}(2 - X_A)}{(1 - 0.3X_A)}$
C	0	$\frac{9.14 \, x \, 10^{-3}X_A}{(1 - 0.3X_A)}$
I	$\frac{0.1 \, x \, 18.28}{600} = 0.003047$	$\frac{3.047 \, x \, 10^{-3}}{(1 - 0.3X_A)}$
Totals	0.03047	$\frac{0.03047 - 9.14 \, x \, 10^{-3}X_A}{(1 - 0.3X_A)}$

Example 4.10

It is desired to carry out the reaction

$$A \rightarrow Products$$

in a tubular reactor. The reaction is first order and occurs in the gas phase without volume change. A previous experiment showed that the first order rate constant was 0.001 min^{-1} at 273 K and 0.05 min^{-1} at 373 K. If 600 m^3/h of pure A is to be treated at 353 K and at 1 atm, determine the volume of the plug flow reactor required to achieve 60 % conversion of A. Assume that the universal gas constant, R = 8.314 kJ/kmol.K and that 1 kmol of an ideal gas occupies 22.41 m^3 at STP.

Answer

Since the reaction rate constant is given by

$$k = Ae^{-\Delta E/RT}$$

where A is the pre-exponential factor and ΔE is the activation energy for the reaction, we can see that,

at 273 K

$$k_{273} = 0.001 = Ae^{-\Delta E/273\,R} \qquad (1)$$

at 373 K

$$k_{373} = 0.05 = Ae^{-\Delta E/373\,R} \qquad (2)$$

From (1) and (2)

$$\frac{0.001}{0.05} = \exp\left[-\frac{\Delta E}{R}\cdot\left(\frac{373-273}{373\ x\ 273}\right)\right]$$

That is

$$0.02 = \exp(-1.1812\ x\ 10^{-4}\Delta E)$$

or

$$\Delta E = \frac{-3.9120}{-1.1812\ x\ 10^{-4}} = 33118.86, kJ/kmol \qquad (3)$$

Substituting (3) in (1)

$$A = \frac{0.001}{e^{-33118.86/(273\ x8.314)}} = \frac{0.001}{4.6021\ x\ 10^{-7}} = 2172.92 \qquad (4)$$

Thus the reaction rate constant for this reaction is

$$k = 2172.92e^{-33118.86/8.314T} = 2172.92e^{-3983.5/T} \qquad (5)$$

Since the reaction is first order, the reaction rate equation becomes

$$-r_A = kC_A = kC_{Ao}(1 - X_A) \qquad (6)$$

where

$$k = 2172.92e^{-3983.5/T} \qquad from\ (5)$$

At 353 K

$$k = 2172.92e^{-3983.5/353} = 2172.92\ x\ 1.2564\ x\ 10^{-5}$$
$$= 0.0273\ min^{-1} \qquad (7)$$

For a plug flow reactor

$$V = \int_0^{X_A} \frac{F_{Ao}dX_A}{-r_A} = \int_0^{X_A} \frac{F_{Ao}dX_A}{kC_{Ao}(1 - X_A)}$$
$$= \frac{V_o}{k}\int_0^{X_A} \frac{dX_A}{(1 - X_A)} = \frac{V_o}{k}\ln\frac{1}{(1 - X_A)} \qquad (8)$$

since $\frac{F_{Ao}}{C_{Ao}} = V_o$ and k is constant during reaction at constant temperature.

Substituting the given values

$$V = \frac{V_o}{k}\ln\frac{1}{(1 - X_A)} = \frac{600}{60\ x\ 0.0273}\ln\frac{1}{(1 - 0.6)}, \frac{m^3}{h}\frac{h}{min}\frac{min}{1}$$
$$= 335.64\ m^3\ \ Ans$$

Example 4.11

At 600 K, the elementary gas phase reaction

$$C_2H_4 + Br_2 \underset{\leftarrow\ k_2}{\overset{k_1\ \rightarrow}{\text{———}}} C_2H_4Br_2$$

has rate constants k_1 = 500 *l/mol.h*, k_2=0.032 h^{-1}. A plug flow reactor is to be fed 600 m^3/h of gas containing 60 % Br_2 and 30 % C_2H_4 and 10 % inerts by volume at 600 K and 1.5 atm. Determine

a). The concentration of C_2H_4, Br_2, and $C_2H_4\ Br_2$ at any conversion X of C_2H_4

b). The equilibrium constant of the reaction at 600 K and 1.5 atm
c). The equilibrium conversion of C_2H_4 at this temperature

Note: 1 kmol of any ideal gas at S.T.P. occupies 22.41 m^3.

Answer

Take C_2H_4 to be the key reactant, A, for the reaction

$$C_2H_4 + Br_2 \underset{\leftarrow k_2}{\overset{k_1 \rightarrow}{\text{——}}} C_2H_4Br_2$$

We can construct a stoichiometric table for the reaction as follows

Basis: 1 kmol of initial reaction mixture

Specie	**Moles initially present**	**Moles reacted**	**Moles remaining**
C_2H_4 (A)	0.30	- $0.30X_A$	$0.30(1\text{-}X_A)$
Br_2 (B)	0.60	- $0.30X_A$	$0.30(2\text{-}X_A)$
C_2H_4 Br_2 (C)	0	$0.30X_A$	$0.30X_A$
Inerts (I)	0.10	-	0.10
	1.00		$1.0\text{-}0.30X_A$

The volume of the reaction mixture is given by

$$V = V_o(1 - \epsilon X_A) \tag{1}$$

where

$$\epsilon = \delta y_{A_o} \tag{2}$$

δ = change in stoichiometric coefficient, and y_{Ao} is the initial mole fraction of reactant A in the initial reaction mixture of initial volume V_o..

From the data given

$$\delta = \frac{1-1-1}{1} = -1 \tag{3}$$

$$y_{A_o} = -0.3 \tag{4}$$

$$\epsilon = \delta y_{A_o} = -0.3 \tag{5}$$

For F_A kmols of reacting gas flowing, the concentration, as kmols per unit volume, of the reactants and products, become for

A	C_2H_4	$\frac{0.30F_A}{V_o}\left(\frac{1-X_A}{1-0.3X_A}\right)$
B	Br_2	$\frac{0.30F_A}{V_o}\left(\frac{2-X_A}{1-0.3X_A}\right)$
C	$C_2H_4\ Br_2$	$\frac{0.30F_AX_A}{V_o(1-0.3X_A)}$
I	Inerts	$\frac{0.10F_A}{V_o(1-0.3X_A)}$

The number of kmols, F_A, fed into the reactor, is obtained using $PV = F_ART$ from which

$$F_A = \frac{P}{P_{STD}}\ x\ \frac{T_{STD}}{T}\ x\ \frac{V}{V_{STD}}\ x\ N_{STD} = \frac{1.5}{1}\ x\ \frac{273}{600}\ x\ \frac{600}{22.41}\ x\ 1, \frac{m^3}{h}\cdot\frac{kmol}{m^3}$$

$$= 18.27\ \frac{kmol}{h} \tag{6}$$

where T_{STD}, P_{STD}, V_{STD}, $N_{STD} = 1$, are the standard temperature, pressure, volume and moles at STP.
This gives

$$\frac{0.30F_A}{V_o} = \frac{0.30\ x\ 18.27}{600}, \frac{kmol}{h}\frac{h}{m^3} = 9.14\ x\ 10^{-3}\frac{kmol}{m^3} \tag{7}$$

Hence the concentrations at any conversion, X_A, are

C_2H_4 (A)	$9.14\,x\,10^{-3}\left(\frac{1-X_A}{1-0.3X_A}\right)$
Br_2 (B)	$9.14\,x\,10^{-3}\left(\frac{2-X_A}{1-0.3X_A}\right)$
$C_2H_4\,Br_2$ (C)	$\frac{9.14\,x\,10^{-3}X_A}{(1-0.3X_A)}$
Inerts (I)	$\frac{3.05\,x\,10^{-3}}{(1-0.3X_A)}$

b). The equilibrium constant, K_C, is then

$$K_C=\frac{k_1}{k_2}=\frac{500}{0.032},\frac{l}{mol\,.\,h}\frac{h}{1}=15{,}625\ m^3/kmol \qquad Ans$$

c) Since

$$K_C=\frac{C_{C_2H_4Br_2}}{C_{C_2H_4}\,.\,C_{Br_2}} \tag{8}$$

Then, at equilibrium, $X_A = X_{AE}$ and

$$15625=\frac{\frac{9.14\,x\,10^{-3}X_{AE}}{(1-0.3X_{AE})}}{9.14\,x\,10^{-3}\left(\frac{1-X_{AE}}{1-0.3X_{AE}}\right)\,x\,9.14\,x\,10^{-3}\left(\frac{2-X_{AE}}{1-0.3X_{AE}}\right)}$$

from which

$$15625=\frac{X_{AE}(1-0.3X_{AE})}{9.14\,x\,10^{-3}(1-X_{AE})(2-X_{AE})}=\frac{X_{AE}-0.3X_{AE}{}^2}{9.14\,x\,10^{-3}(2-3X_{AE}+X_{AE}{}^2)}$$

or $$142.81(2-3X_{AE}+X_{AE}{}^2)=X_{AE}-0.3X_{AE}{}^2$$

That is $$143.11X_{AE}{}^2-429.43X_{AE}+285.62=0 \tag{9}$$

By Pythagoras theorem

$$X_{AE}=\frac{+429.43\pm\sqrt{184410.12-163500.31}}{286.22}=\frac{429.43\pm144.60}{286.22}$$
$$=0.995\ or\ 2.005$$

It is clear that $X_{AE} = 2.005$ is not possible as fractional conversion cannot be greater than 1. Thus the answer is $X_{AE} = 0.995$

Example 4.12

Define the terms (a) space time (b) space velocity with respect to flow reactors. What is their industrial significance?

The reaction

$$N_2 + O_2 \rightleftharpoons 2NO$$

takes place in a CSTR. The rate of reaction, r, at about 2700 C, is related to the fractional conversion of nitrogen, X, as

$$r = k_1 C_o^2[(1 - X)(0.195 - X) - 400X^2]$$

where C_o is the inlet nitrogen concentration.

If the inlet concentration of nitrogen is 6.95 x 10^{-2} kmol/m^3, k_1, the forward rate constant, and the volumetric flow rate into the reactor is 20 m^3 per second, determine the volume of the reactor required to achieve 80 % of the equilibrium conversion of 20 %.

Answer

Space time is the time required to feed into the reactor an amount of material equivalent to one reactor volume at entrance conditions. It is expressed mathematically as

$$\tau = \frac{V}{u_o}$$

where V is the reactor volume at any time during reaction and u_o is the volumetric flow into the reactor at initial conditions.

Space velocity, on the other hand, is the inverse of space time and is given by

$$space\ velocity = \frac{1}{\tau} = \frac{u_o}{V}$$

The industrial significance of both terms is that they give an approximate, and often necessarily useful, estimate of processing times or rates of particular reaction feed stocks.

For a CSTR

$$V = \frac{F_{OUT} - F_{IN}}{r} = \frac{F_{Ao}X_A}{r_A} \tag{1}$$

At equilibrium

$$r_{Ae} = k_1 C_o^2\left[(1 - X_{Ae})(0.195 - X_{Ae}) - 400{X_{Ae}}^2\right] = 0$$

where r_{Ae} and X_{Ae} are the values at equilibrium. Thus

$$(1 - X_{Ae})(0.195 - X_{Ae}) = 400{X_{Ae}}^2$$

That is

$$0.195 - 0.195X_{Ae} - X_{Ae} + {X_{Ae}}^2 = 400{X_{Ae}}^2$$

from which

$$399{X_{Ae}}^2 + 1.195X_{Ae} - 0.195 = 0$$

This has the Pythagoras equation solution of

$$X_{Ae} = \frac{-1.195 \pm \sqrt{(1.195)^2 + 4\,x\,0.195\,x\,399}}{2\,x\,399}$$

$$= \frac{-1.195 \pm 17.6819}{798} = 0.02$$

$$X_A = 80\,\%\,of\,X_{Ae} = 0.8\,x\,0.20 = 0.016 \tag{3}$$

$$F_{Ao} = u_{Ao}C_{Ao} = 20\,\frac{m^3}{s}\,x\,6.95\,x\,10^{-2}\frac{kmol}{m^3} = 1.39\,\frac{kmol}{s} \tag{4}$$

The rate of reaction is

$$r_A = k_1 C_{Ao}^2\left[(1 - X_A)(0.195 - X_A) - 400{X_A}^2\right]$$

$$= k_1(6.95\,x\,10^{-2})^2[(1 - 0.016)(0.195 - 0.016) - 400(0.016)^2]$$

$$= 3.5616\,x\,10^{-4}k_1 \tag{5}$$

Substituting (3) and (4) in (1)

$$V = \frac{F_{Ao}X_A}{r_A} = \frac{1.39 \ x \ 0.016}{3.5616 \ x \ 10^{-4}k_1} = \frac{62.44}{k_1}, m^3 \quad Ans$$

Example 4.13

Outline, with examples, the methods of analysis applied to multiple reactor systems for the situations in which

a). Different types of reactors are combined
b). One reactor type is used but there is need to determine either the number or size required to achieve a given fractional conversion or the fractional conversion achieved for a given number or size of the same type of reactor.

Answer

When different types of reactors are combined

Only the combination of the plug flow and continuous stirred tank reactors may be considered for the practical reason that they are both flow reactors. Combining a batch reactor with either or both of these flow reactors would be impractical because of their different operating time scales.

One possible method of combining flow reactors is to have them operate in series. In this scheme, we can put the CSTR before the PFR or the PFR before the CSTR. In most cases, the number of PFR and CSTR to be combined are known.

a): ***One CSTR before one PFR in series***

This is shown schematically as follows

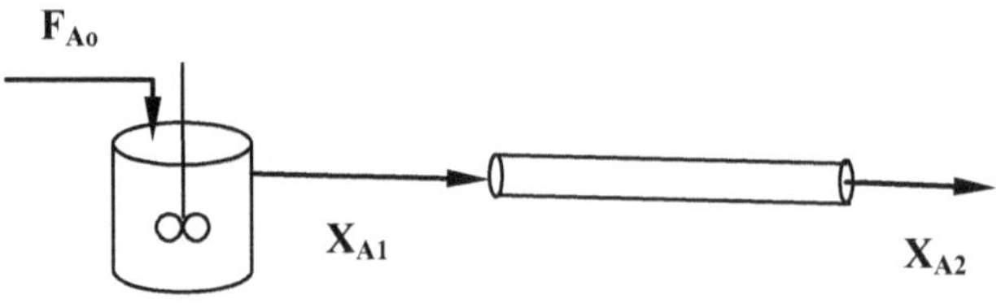

The material balance equations are, for the CSTR

$$V_{CSTR} = \frac{F_{Ao}X_{A1}}{-r_A}$$

and for the PFR

$$V_{PFR} = \int_{X_{A1}}^{X_{A2}} \frac{F_{A1}dX_A}{-r_A}$$

The reaction rate – conversion plot is then as follows

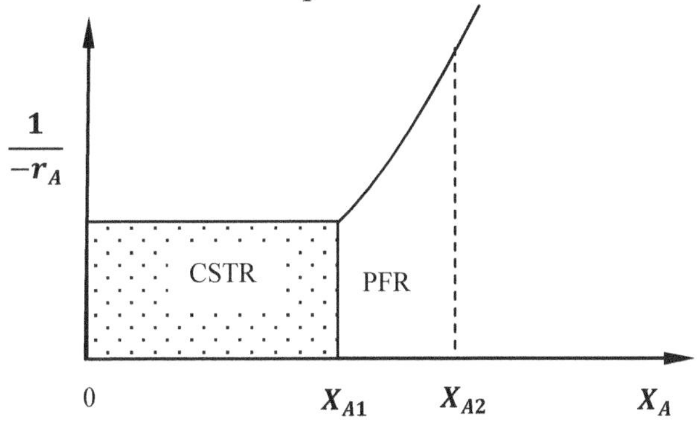

b): ***One PFR before one CSTR in series***

This is shown schematically as follows

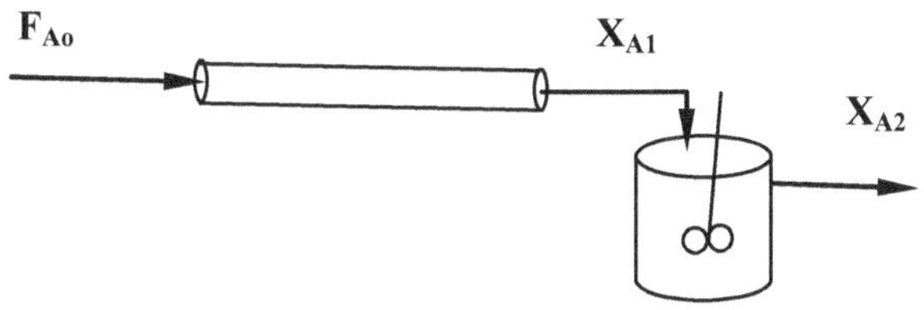

The material balance equations are, for the CSTR,

$$V_{CSTR} = \frac{F_{Ao}X_{A1}}{-r_A}$$

and for the PFR,

$$V_{PFR} = \int_{X_{A1}}^{X_{A2}} \frac{F_{A1}dX_A}{-r_A}$$

The reaction rate – conversion plot is then as follows

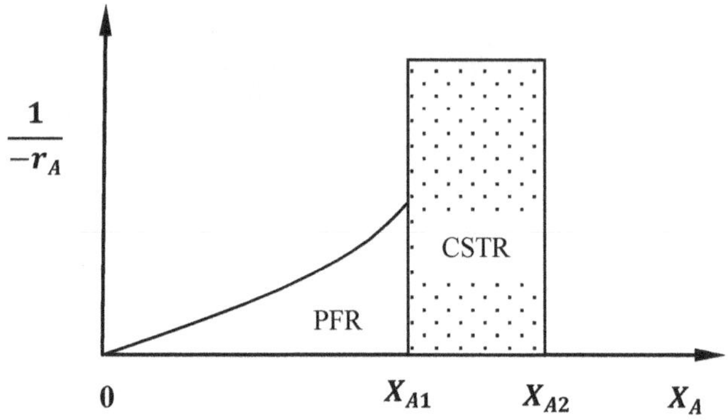

In both cases, it is the capital cost, operational cost and maintenance cost which dictate which reactor volume should be reduced which enlarged to achieve the given conversion.

b): ***PFR and the CSTR in parallel***

This is a trivial solution and not really a combination, as both reactors have to achieve the same conversion. The possible benefit is that of increased processing rate as determined by the relative sizes of the two reactors. A possible disadvantage is that the products of the two different reactors may not have the same quality as a consequence of their different modes of processing. In series combination, whatever difference was introduced by the different processing style was smoothed out in the final reactor

<u>When the same types of reactors are combined</u>

In this situation, we can have either a number of PFR in series or a number of CSTR in series.

b): ***PFR in series***

This is shown, schematically, as follows

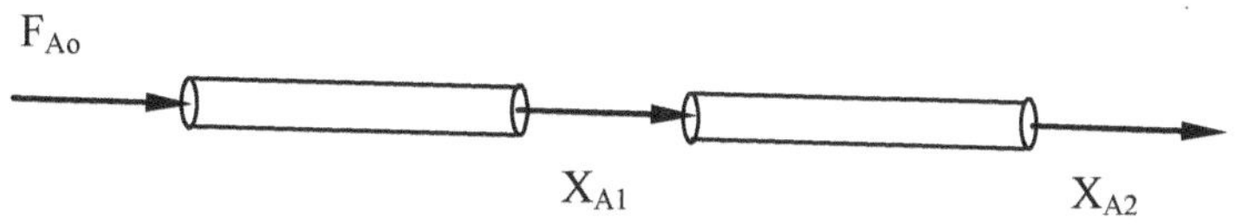

The material balance equations are

$$V_{PFR} = \int_{X_{Ao}}^{X_{A1}} \frac{F_{Ao} dX_A}{-r_A} + \int_{X_{A1}}^{X_{A2}} \frac{F_{A1} dX_A}{-r_A} = \int_{X_{Ao}}^{X_{A2}} \frac{F_{Ao} dX_A}{-r_A}$$

The reaction rate – conversion plot is then as follows

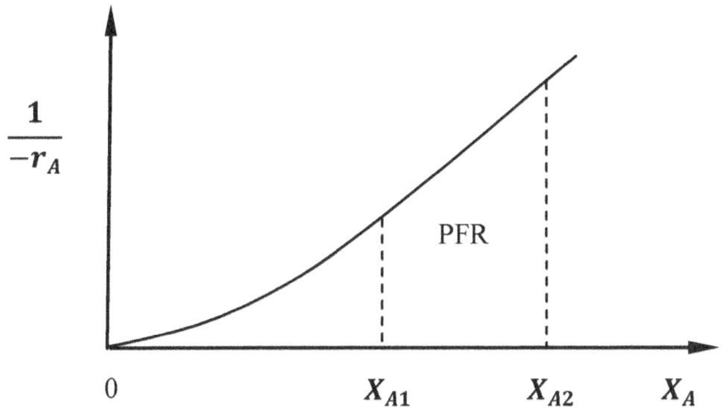

This plot shows that there is no saving in reactor volume or size for a given conversion. It is also easy to determine, from the graph, the number, size of reactors or conversion for any series combinations of plug flow reactors

b): **CST*R in series***

The schematics is shown below

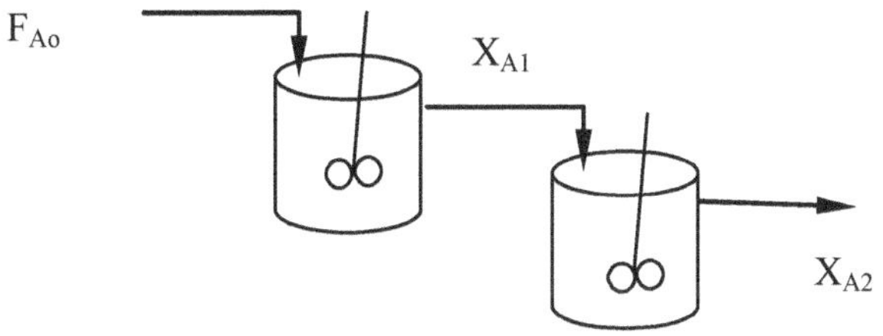

If conversion and the number of reactors are known, the material balance equations are, for two CSTRs,

$$V_{CSTR} = \frac{F_{Ao}X_{A1}}{-r_A} + \frac{F_{A1}X_{A2}}{-r_A}$$

The reaction rate – conversion plot is then as follows

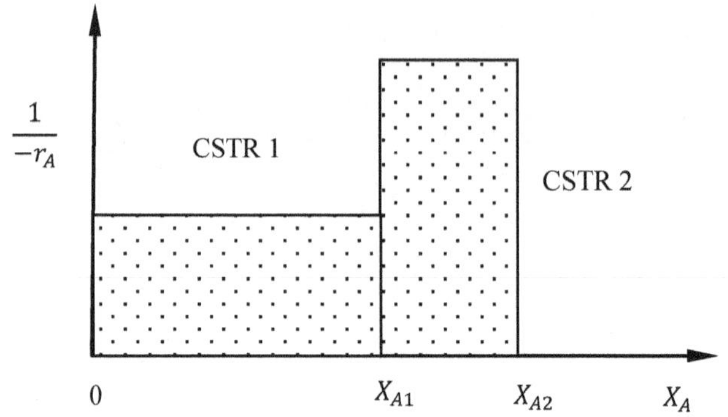

If the volume of the CSTR is known but the conversion or number is to be determined, the method of analysis is to resort to the primary material balance equations. That is

$$F_{Ao} = F_{A1} + r_A V_1$$

$$F_{A1} = F_{A2} + r_A V_2$$

....

$$F_{A,n-1} = F_{An} + r_A V_n$$

The last and generalised equation is the same as

$$u_{n-1}C_{n-1} = u_n C_n + r_n V_n$$

from which

$$r_n = \frac{u_{n-1}}{V_n}C_{n-1} - \frac{u_n}{V_n}C_n$$

If $u_{n-1} = u_n$,

$$\frac{u_{n-1}}{V_n} = \frac{u_n}{V_n} = \frac{1}{\theta_n}$$

$$r_n = \frac{1}{\theta_n}C_{n-1} - \frac{1}{\theta_n}C_n$$

The conversion and/or number of CSTR can now be determined by successive calculation or from a graph of $-r_A$ versus C_A. as shown below. If conversion, X_{An}, is desired to be determined, it can be read off from this graph. If X_{An} is known but $\boldsymbol{n}$ is to be determined, a trial and error numerical or graphical procedure is used.

Determining Conversion or Number of CSTRs Graphically (Walas, 1959)

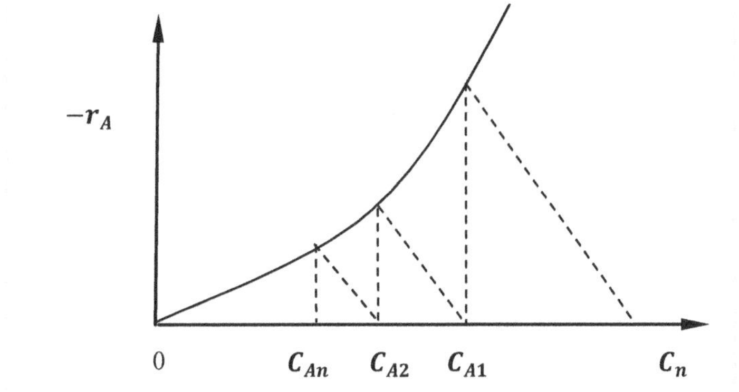

Example 4.14

1000 litres/h of radioactive fluid, having a half life of 20 h, is treated by passing it through two ideal stirred tanks in series, each of volume 40,000 litres. How much has the activity of the fluid decayed in passing through this system?

Answer

The system may be represented, schematically, as follows

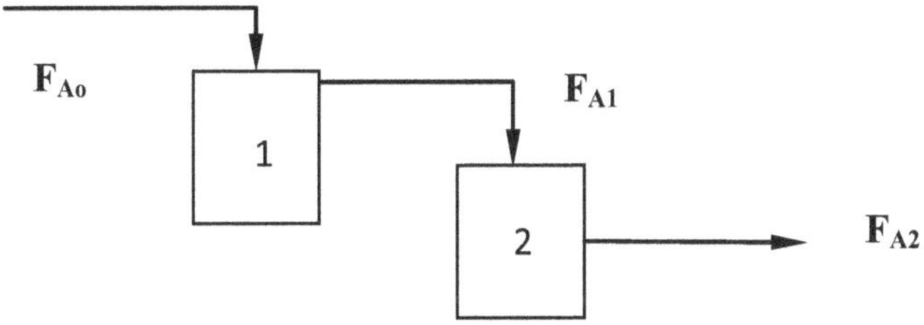

For a first order reaction, the fractional conversion of key reactant, A, is given by

$$X_A = 1 - e^{-k_1 t} \tag{1}$$

where k_1 is the first order rate constant.

The half life, $\tau_{1/2}$, of a first order reaction can, also, be shown to be

$$\tau_{½} = \frac{\ln 2}{k_1} \tag{2}$$

From the data given and equation (2)

$$k_1 = \frac{\ln 2}{\tau_{½}} = \frac{\ln 2}{20} = 0.03466\, h^{-1} \tag{3}$$

Since the two tanks have equal volumes, the time spent by the radioactive fluid in each tank is the same and is given by

$$t = \frac{40{,}000}{1000}, \frac{litres}{1} \frac{h}{litres} = 40\, h \tag{4}$$

From equation (1) with the radioactive fluid passing through the first tank

$$X_{A1} = 1 - e^{-0.03466\, x\, 40} = 1 - 0.25 = 0.75 \tag{5}$$

Similarly, for the second tank

$$X_{A2} = 1 - e^{-0.03466\, x\, 40} = 1 - 0.25 = 0.75 \quad (6$$

If the activity of the radioactive fluid entering the first tank is denoted by C_{Ao}, then

$$C_{A1} = C_{Ao}(1 - X_{A1}) = 0.25C_{Ao} \tag{7}$$

The activity of the radioactive fluid leaving the second tank becomes

$$C_{A2} = C_{A1}(1 - X_{A2}) = 0.25C_{A1} = 0.25\, x\, 0.25C_{Ao} = 0.0625C_{Ao} \tag{8}$$

Thus the activity has decayed 1- 0.0625 = 0.9375 of 93.75 % Ans

Example 4.15

A solution containing 8.0 kmol/m^3 of a reactive component is to be treated at the rate of 0.708 m^3/h. Kinetic data obtained in a batch laboratory reactor are as follows

C_A, kmol/m^3	r_A, kmol/h.m^3
8.025	13.64
7.223	10.83
6.420	8.510
5.620	6.500
4.820	4.980
4.010	3.850
3.210	2.890
2.410	2.050
1.610	1.300
0.800	0.640

a. If the filling and draining time per batch is negligible, what size of batch reactor is needed for 90 % conversion? How many batches are made in a 24-hour day?
b. What percentage conversion is attained with a two stage reactor battery, each vessel being 1.42 m^3?
c. For 90 % conversion, determine the total reactor volumes required with one stage and two stages.

Answer

a. For any chemical reactor

$$F_{Ao} = F_A + r_A V + \frac{dN_A}{dt}$$

where F_{Ao}, F_A are the flow into, and flow out of, respectively, the reactor, r_A the rate of chemical reaction, V the volume of the reaction mixture and N_A the number of moles of reacting substance.

For a batch reactor, $F_{Ao} = F_A = 0$ so that

$$-r_A = \frac{1}{V}\frac{dN_A}{dt}$$

When V is constant

$$-r_A = \frac{dC_A}{dt}$$

Since $C_A = C_{Ao}(1 - X_A)$

$$r_A = C_{Ao}\frac{dX_A}{dt}$$

Thus, the time required for the concentration to change from C_{Ao} to C_A, in the reactor, is given by

$$t = C_{Ao}\int_{C_{Ao}}^{C_A}\frac{dX_A}{r_A} = \frac{N_{Ao}}{V_o}\int_{C_{Ao}}^{C_A}\frac{dX_A}{r_A} = \int_{C_{Ao}}^{C_A}\frac{dC_A}{-r_A} \qquad (1)$$

For 90 % conversion, $X_A = 0.9$ and $C_{Ao} = 8.0\ kmol/m^3$

$$C_A = C_{Ao}(1 - X_A) = 8(1 - 0.9) = 0.8\ kmol/m^3$$

We can see from equation (1) that integration of equation (1) within the limits of C_A and C_{Ao} will give the required value of t. Because the values of r_A are given as a set of data, this integration will have to be either numerical or graphical (of course one can fit the r_A versus C_A data to an equation form and then integrate equation (1) analytically).

The given data is plotted to get insight into the relationship between r_A and C_A. This is shown in Fig. 4.15a below.

Fig. 4.15a indicates that the order of the reaction may be between

1 and 2. To get a graph of $1/-r_A$ versus C_A, the given data are recalculated and tabulated as shown in Table 4.15a below

A graphical plot of $1/r_A$ versus C_A is shown in Fig. 4.15b from which the area under the curve gives the desired value of t.

Alternatively, the area under the curve may be found by either Simpson's or Weddle's rule

Fig. 4.15a: Plot of $-r_A$ versus C_A

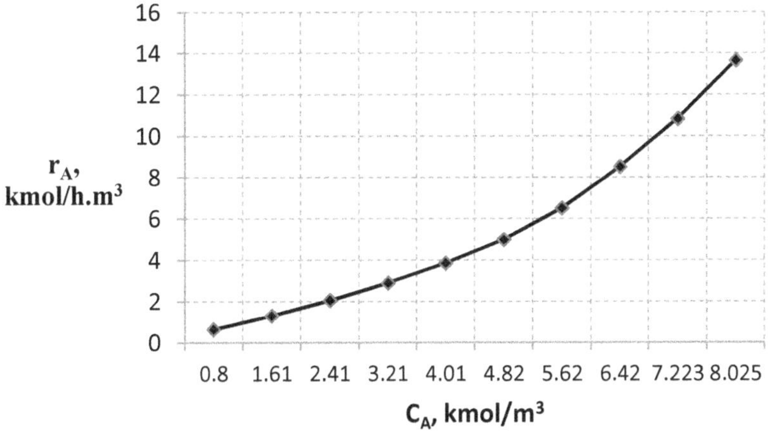

Table 4.15a; Table of $1/-r_A$ versus C_A

C_A, kmol/m³	$-r_A$, kmol/h.m³	$1/-r_A$	C_A, kmol/m³
0.8	0.64	1.563	0.800
1.61	1.3	0.769	1.610
2.41	2.05	0.488	2.410
3.21	2.89	0.346	3.210
4.01	3.85	0.260	4.010
4.82	4.98	0.201	4.820
5.62	6.5	0.154	5.620
6.42	8.51	0.118	6.420
7.223	10.83	0.092	7.223
8.025	13.64	0.073	8.025

To use Weddle's rule, the interval ***h*** is given by six elements defined by seven x-ordinate values. Thus

$$h = \frac{(8-0.8)}{6} = 1.2\ kmol/m^3 \qquad (2)$$

Fig. 4.15b: Plot of 1/-r$_A$ versus C$_A$

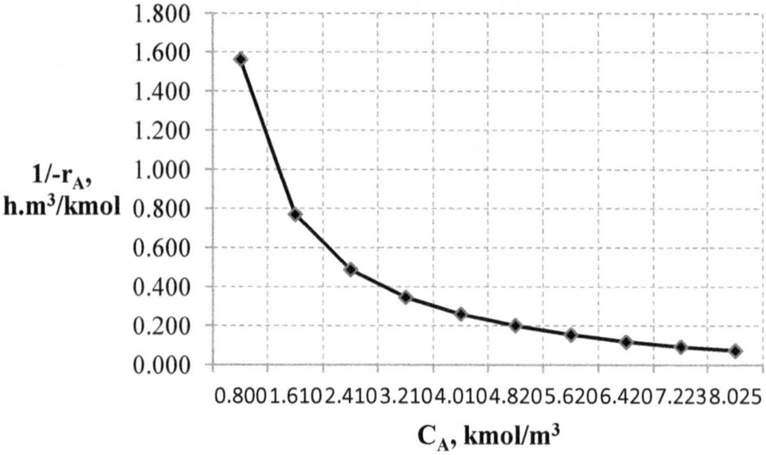

Table 4.15b lists the evenly spaced six values of C_A and $-r_A$ read of from Fig. 4.15a at these intervals

Table 4.15b; Table of 1/-r$_A$ at evenly spaced values of C$_A$

C_A, kmol/m^3	1/-r_A, h.m^3/kmol	f designation
0.8	1.563	f_1
2.0	0.61	f_2
3.2	0.35	f_3
4.4	0.22	f_4
5.6	0.15	f_5
6.8	0.105	f_6
8.0	0.07	F_7

By Weddle's rule

$$\int_{C_{Ao}}^{C_A} \frac{dC_A}{-r_A} = \frac{3}{10} h(f_1 + 5f_2 + f_3 + 6f_4 + f_5 + 5f_6 + f_7) \qquad (3)$$

That is

$$\int_{C_{Ao}}^{C_A} \frac{dC_A}{-r_A} = \frac{3}{10} x\ 1.2\ (1.563 + 5\ x\ 0.61 + 0.35 + 6\ x\ 0.22 + 0.15$$
$$+ 5\ x\ 0.105 + 0.07) = 2.530\ h$$

Since one batch takes 2.530 hours, there will be 24/2.530 = 9.49 or 10 batches per day. Ans.

b. If the solution is to be treated in a two stage reactor battery at the rate of 0.708 m^3/h, it will be safe to assume that this will be done in two CSTRs in series. Thus we can illustrate the schematics of the operation as follows

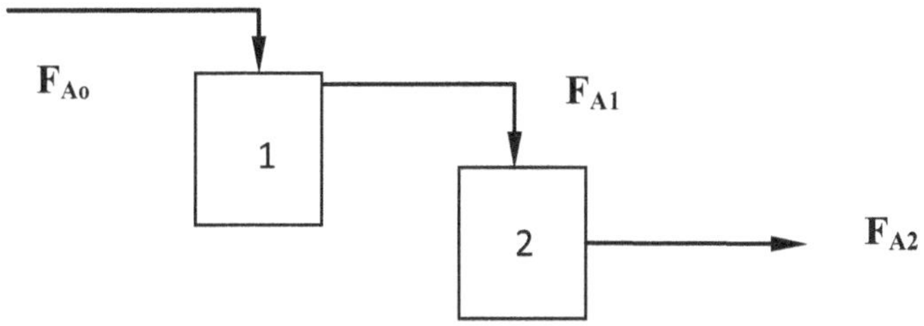

For two CSTR in series of equal volume, V

$$F_{Ao} = F_{A1} + r_{A1}V \qquad (1a)$$

and

$$F_{A1} = F_{A2} + r_{A2}V \qquad (1b)$$

But equation (1a) is equivalent to

$$u_o C_{Ao} = u_o C_{A1} + r_{A1} V$$

That is

$$C_{Ao} = C_{A1} + r_{A1} \frac{V}{u_o} = C_{Ao} = C_{A1} + r_{A1}\,\theta$$

from which

$$r_{A1} = \frac{C_{Ao}}{\theta} - \frac{C_{A1}}{\theta} \qquad (2a)$$

and from equation (1b)

$$r_{A2} = \frac{C_{A1}}{\theta} - \frac{C_{A2}}{\theta} \qquad (2b)$$

where

$$\frac{V}{u_o} = \theta = \frac{1.42}{0.708}, \frac{m^3}{1} \frac{h}{m^3} = 2.006\,h \tag{3}$$

From (2a), (3) and $C_{Ao} = 8.0$ kmol/m^3,

$$r_{A1} = 3.998 - 0.499\,C_{A1} \tag{4a}$$

If we calculate and list equation (4a) in Table 4.15c and plot it into Fig 4.15a, we shall get its solution at the intersection of equation (4a) with the $-r_A$ versus C_A curve given as shown in Fig. 4.15c. Thus

Table 4.15c: Tabulation of Equation (4a)

C_A, kmol/m^3	$-r_A$, kmol/h.m^3	$-r_{A1}$, kmol/h.m^3
0.8	0.64	3.60
1.6	1.3	3.20
2.4	2.05	2.80
3.2	2.89	2.40
4	3.82	2.00
4.8	4.84	1.60
5.6	5.95	1.20
6.4	7.15	0.80
7.2	8.44	0.41
8	9.82	0.01

Fig. 4.15c: Determination of $-r_{A1}$ and C_{A1}

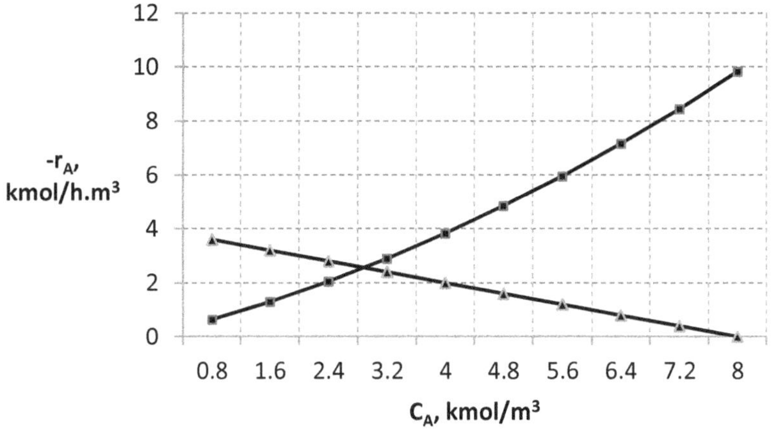

This intersection occurs at $C_{A1} = 2.884$ kmol/m^3 and $-r_{A1} = 2.558$ kmol/h.m^3. From (2b), (3) and $C_{A1} = 2.884$ kmol/m^3,

$$r_{A2} = 1.4377 - 0.499\, C_{A2} \qquad (4b)$$

If, again, we calculate and list equation (4b) in Table 4.15d and plot it into Fig 4.15a, we shall get its solution at the intersection of equation (4b) with the $-r_A$ versus C_A curve given as shown in Fig. 4.15d. Thus

Table 4.15d: Tabulation of Equation (4b)

C_A, kmol/m^3	$-r_A$, kmol/h.m^3	$-r_{A2}$, kmol/h.m^3
0.8	0.64	1.04
1.6	1.3	0.64
2.4	2.05	0.24
3.2	2.89	-0.16
4	3.82	-0.56
4.8	4.84	-0.96
5.6	5.95	-1.36
6.4	7.15	-1.76
7.2	8.44	-2.16
8	9.82	-2.55

Fig. 4.15d: Determination of r_{A2} and C_{A2}

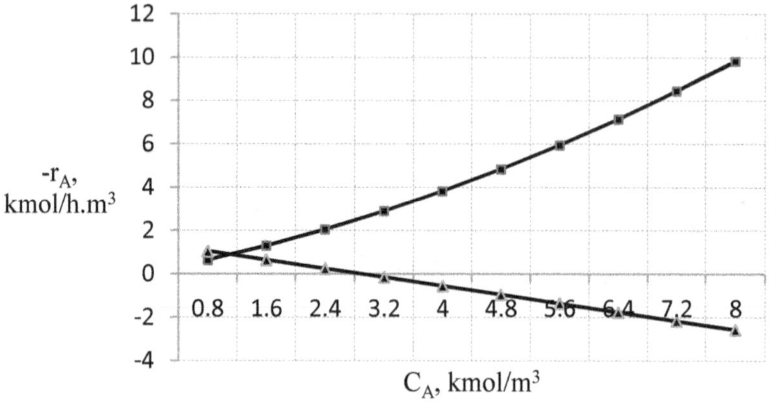

This intersection occurs at $C_{A2} = 1.101$ kmol/m^3 and $-r_{A2} = 0.888$ kmol/h.m^3.

The fractional conversion in the first reactor is

$$X_{A1} = \frac{C_{Ao} - C_{A1}}{C_{Ao}} = \frac{8.0 - 2.884}{8.0} = 0.6395 \qquad (5)$$

That in the second reactor is

$$X_{A2} = \frac{C_{A1} - C_{A2}}{C_{A1}} = \frac{2.884 - 1.101}{2.884} = 0.6182 \qquad (6)$$

The overall fractional conversion is

$$X_A = \frac{C_{Ao} - C_{A2}}{C_{Ao}} = \frac{8.0 - 1.101}{8.0} = 0.8624 \quad or \quad 86.24\,\% \quad Ans$$

c. For a total conversion of 90%, assuming this time, that all the reactors under consideration are CSTR,

<u>For a single CSTR</u>, equation (2a) gives, since $C_{Ao} = 0.8\ kmol/m^3$,

$$r_{A1} = \frac{C_{Ao}}{\theta} - \frac{C_{A1}}{\theta} = \frac{8}{\theta} - \frac{1}{\theta}C_{A1} \qquad (7)$$

This is the equation of a straight line, of slope $-\frac{1}{\theta}$.

The solution will be found where this line, with

$$C_A = C_{Ao}(1 - X_A) = 8.0\ x\ 0.1 = 0.8\ kmol/m^3$$

intersects the given $-r_A$ versus C_A curve. This occurs at $-r_A = 0.64$. Thus from (7)

$$r_{A1} = \frac{8}{\theta} - \frac{0.8}{\theta} = \frac{7.2}{\theta} = 0.64 \quad or \quad \theta = 11.25$$

From (3)

$$\frac{V}{u_o} = \theta, giving \;\; V = 0.708\ x\ 11.25, \frac{m^3}{h}\frac{h}{1} = 7.965\ m^3 \;\; Ans.$$

<u>For two stages or two CSTRs</u>, equation (2a) gives, since $C_{Ao} = 0.8\ kmol/m^3$,

$$r_{A1} = \frac{C_{Ao}}{\theta} - \frac{C_{A1}}{\theta} = \frac{8}{\theta} - \frac{1}{\theta}C_{A1} \qquad (8)$$

and

$$r_{A2} = \frac{C_{A1}}{\theta} - \frac{C_{A2}}{\theta} = \frac{C_{A1}}{\theta} - \frac{0.8}{\theta} \qquad (9)$$

Thus, we have two equations with four unknowns. The only way this problem can be handled is by a trial and error procedure, either graphically or analytically.

A graphical procedure is easier, in this case, because we have been given the numerical data relating $-r_A$ and C_A. The method used in Figures 4.15c and 4.15d will be applied.

It consists of drawing a straight line of slope $-\frac{1}{\theta_1}$ from the point (C_{Ao}, $-r_{Ao}$) to cut the $-r_A$ versus C_A curve at the point (C_{A1}, $-r_{A1}$), dropping vertically down to the C_A axis to C_{A1} and then drawing another straight line of slope $-\frac{1}{\theta_2}$ to cut the $-r_A$ versus C_A curve at the point (C_{A2}, $-r_{A2}$).

If desired, $-\frac{1}{\theta_1}$ can be chosen to be equal to $-\frac{1}{\theta_2}$. In all cases, it is still a trial and error procedure as the slope has to be manipulated in order for equations (8) and (9) to be satisfied.

Figures 4.15e and 4.15f demonstrate the two cases in which $-\frac{1}{\theta_1}$ is chosen to be different from $-\frac{1}{\theta_2}$ and in which $-\frac{1}{\theta_1}$ is chosen to be equal to $-\frac{1}{\theta_2}$.

In the situation in which $\theta_1 \neq \theta_2$, we have the freedom to choose θ_1 at random and then find a $\theta_2$that makes both values satisfy equations (8) and (9).

If we choose C_{A1} = 3.0 kmol/m^3, we shall find that the straight line, based on equation (8), which intersects the curve of the original data at C_{A1} = 3.0 kmol/m^3, has a slope of 0.509 or a value of $\theta_1 = 1.9646$. Thus, from equation (3)

$$V_1 = \theta_1 u_o = 1.9646\ x\ 0.708, \frac{h\,m^3}{1\ h}$$
$$= 1.3909\ m^3 \cong 1.39\ m^3\ Ans$$

All we have to do now is to start from C_{A1} = 3.0 kmol/m^3 at the C_A axis and draw a straight line to intersect the curve of the original $-r_A$ versus C_A data at $-r_A$ = 0.64 and C_{A2} = 0.8.

This straight line has to satisfy equation (9), for which $-r_{A2}$ = 0.64 when C_{A2} = 0.8 kmol/m^3. Thus

$$r_{A2} = \frac{3}{\theta} - \frac{0.8}{\theta} = 0.64$$

from which $\theta_2 = 3.4375\ h$. Thus, from equation (3)

$$V_2 = \theta_2 u_o = 3.4375\ x\ 0.708, \frac{h\,m^3}{1\ h} = 2.4338\ m^3$$
$$\cong 1.43\ m^3\ Ans$$

Fig. 4.15e: Determination of θ_1 and θ_2 for two CSTR in series

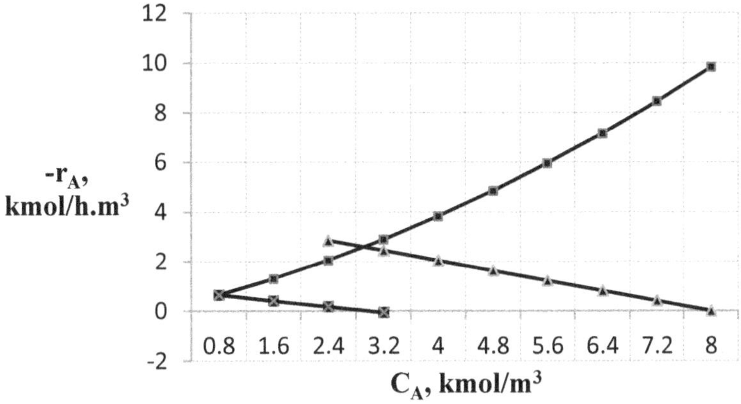

In situation in which $\theta_1 = \theta_2$, equation (9), with $-r_{A2} = 0.64$ when $C_{A2} = 0.8$, shows that

$$r_{A2} = \frac{C_{A1}}{\theta} - \frac{0.8}{\theta} = 0.64 \quad (10)$$

$$r_{A1} = \frac{8}{\theta} - \frac{0.8}{\theta} \quad (11)$$

A trial and error procedure with equation (10) yields $\theta = 2.6829$ from which $C_{A1} = 2.517$ kmol/m^3. Substituting this in equation (11),

$$r_{A1} = \frac{8}{2.6829} - \frac{1}{2.6829} x\, 2.517 = 2.044$$

Thus, from equation (3), the volume of each of the two CSTR is

$$V = \theta u_o = 2.6829\, x\, 0.708, \frac{h\, m^3}{1\ h} = 1.8995\, m^3 \cong 1.9\, m^3 \quad Ans$$

It is easy to see, from the above examples, how the procedure can be extended for any number of CSTRs in series. For operational and equipment maintenance reasons, it is, always, preferable, in the industry, to have equal sized CSTRs. This enables interchange of equipment parts between reactors.

Fig. 4.15f: Determination of -1/θ for Two CSTR in Series when $\theta_1 = \theta_2$

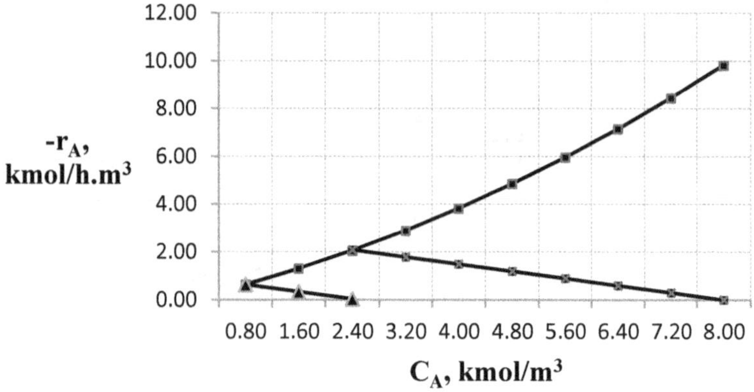

Example 4.16

Styrene is to be made by the catalytic dehydrogenation of ethyl benzene.

$$C_6H_5.CH_2.CH_3 \rightleftharpoons C_6H_5CH : CH_2 + H_2$$

The rate equation for the reaction has been reported as follows

$$-r_A = k\left(P_{Et} - \frac{1}{K_P} P_{St} \, . P_H\right)$$

where $P_{Et,}$ P_{St} and P_H are the partial pressures of ethyl benzene, styrene and hydrogen, respectively.

The reactor will consist of a number of tubes, each of 80 mm internal diameter, packed with catalyst having a bulk density of 1440 kg/m^3. The ethyl benzene will be diluted with steam, the feed rates per unit cross-sectional area being, for ethyl benzene, 1.6 x 10^{-3} kmol/m^2.s; for steam, 29 x 10^{-3} kmol/m^2.s.

The reactor will be operated at an average pressure of 1.2 bar and the temperature will be maintained at 560 C throughout. If the fractional conversion of ethyl benzene is to be 0.45, estimate the length and number of tubes required to produce 20 tonne styrene per day.

At 560 C, k = 6.6 x 10^{-9} kmol . m^2/N. s. kg catalyst. K_P = 1.0 x 10^4 N/m^2 (Fogler, 1974)

Answer

Let us represent ethyl benzene as A, styrene as B and Hydrogen as C. Then the reaction becomes

$$A \rightleftharpoons B + C$$

The volume, V, of a plug flow reactor is given by

$$V = \int_0^{X_A} \frac{F_{Ao} dX_A}{-r_A}$$

We can construct a stoichiometric table for the given reaction as shown below

Specie	Moles initially present	Moles reacted	Moles remaining
Ethyl benzene (A)	F_{Ao}	$-F_{Ao}X_A$	$F_{Ao}(1 - X_A)$
Styrene (B)	F_{Bo}	$F_{Ao}X_A$	$F_{Bo} + F_{Ao}X_A$
Hydrogen (C)	F_{Co}	$F_{Ao}X_A$	$F_{Co} + F_{Ao}X_A$
Inerts (Steam)	F_{Io}		F_{Io}
Total	$F_{Ao} + F_{Bo} + F_{Co} + F_{Io}$		$F_{Ao} + F_{Bo} + F_{Co} + F_{Io} + F_{Ao}X_A$

Molar change on reaction per mole of key reactant

$$\delta = \frac{2 - 1}{1} = 1$$

Initial mole fraction of key reactant

$$y_{Ao} = \frac{1.6\ x\ 10^{-3}}{(1.6 + 29)\ x\ 10^{-3}} = 0.052$$

Fractional volume change of reaction mixture

$$\epsilon = \delta y_{Ao} = 1\ x\ 0.052 = 0.052$$

Volume of reaction mixture

$$V = V_o(1 + \epsilon X_A) = V_o(1 + 0.052X_A)$$

Thus

$$C_A = \frac{F_A}{V} = \frac{F_{Ao}(1 - X_A)}{V_o(1 + 0.052X_A)} = \frac{C_{Ao}(1 - X_A)}{(1 + 0.052X_A)}, \frac{kmol}{m^3}$$

Since $PV = nRT$ and $C = \frac{n}{V} = \frac{P}{RT}$, the partial pressures of the reactants and products become

$$P_A = \frac{P_{Ao}(1 - X_A)}{(1 + 0.052X_A)}, \frac{N}{m^2}$$

$$P_B = \frac{P_{Ao}(\theta_B + X_A)}{(1 + 0.052X_A)}, \frac{N}{m^2}$$

$$P_C = \frac{P_{Ao}(\theta_C + X_A)}{(1 + 0.052X_A)}, \frac{N}{m^2}$$

where

$$\theta_B = \frac{F_{Bo}}{F_{Ao}}, \qquad \theta_C = \frac{F_{Co}}{F_{Ao}}$$

and

$$P_{Ao} = 0.052\ x\ 1.2\ x\ 1.013\ x\ 10^5 = 6321.12\ N/m^2$$

When all these values are substituted into the given equation of reaction, we get

$$-r_A = k\left(\frac{P_{Ao}(1 - X_A)}{(1 + 0.052X_A)} - \frac{1}{K_P} \cdot \frac{P_{Ao}(\theta_B + X_A)}{(1 + 0.052X_A)} \cdot \frac{P_{Ao}(\theta_C + X_A)}{(1 + 0.052X_A)}\right)$$

Since $\theta_B = \theta_C = 0$, k = 6.6 x 10^{-9} kmol . m^2/N. s. kg catalyst. K_P = 1.0 x 10^4 N/m^2, the reaction equation becomes

$$-r_A = 6.6\,x\,10^{-9}\left(\frac{6321.12(1-X_A)}{(1+0.052X_A)} - \frac{1}{1\,x\,10^4}\cdot\frac{6321.12X_A}{(1+0.052X_A)}\cdot\frac{6321.12X_A}{(1+0.052X_A)}\right)$$

$$-r_A = \left(\frac{4.172\,x\,10^{-5}(1-X_A)}{(1+0.052X_A)} - \frac{2.637x\,10^{-5}\,X_A^2}{(1+0.052X_A)^2}\right), \frac{kmol.m^2}{N.s.kg}\cdot\left(\frac{N}{m^2} - \frac{m^2}{N}\cdot\frac{N}{m^2}\cdot\frac{N}{m^2}\right)$$

$$-r_A = \left(\frac{4.172\,x\,10^{-5} - 3.955\,x\,10^{-5}\,X_A - 2.854\,x\,10^{-5}X_A^2}{(1+0.052X_A)^2}\right), \frac{kmol}{s.kg\ catalyst}$$

The volume of the reactor is, therefore,

$$V = \int_0^{X_A} \frac{F_{Ao}dX_A}{-r_A}$$

$$= 10^5 F_{Ao}\int_0^{0.45} \frac{(1+0.052X_A)^2 dX_A}{4.172 - 3.955\,X_A - 2.854X_A^2}, \frac{kmol}{m^2.s}\cdot\frac{s.kg\ catalyst}{kmol}$$

$$= 10^5 F_{Ao}\int_0^{0.45} \frac{(1+0.052X_A)^2 dX_A}{4.172 - 3.955X_A + 2.854\,X_A^2}, \frac{kg\ catalyst}{m^2}$$

The integration will be done numerically using Simpson's rule as this is the most convenient. Thus, a table of X_A versus the integrand is constructed as shown below, with the X_A intervals arranged to suit the number of elements required for Simpson's rule integration.

X_A	Integrand
0	0.239693
0.05625	0.255262

0.1125	0.274113
0.16875	0.297263
0.225	0.326212
0.28125	0.363267
0.3375	0.412168
0.39375	0.479399
0.45	0.577269

Thus, by Simpson's rule,

$$\int_0^{0.45} \frac{(1+0.052X_A)^2 dX_A}{4.172 - 3.955\, X_A - 2.854\, X_A^2},$$

$$= \frac{0.05625}{3}(0.239693 + 4\, x\, 0.255262 + 0.274113)$$

$$+ \frac{0.05625}{3}(0.274113 + 4\, x 0.297263 + 0.326212)$$

$$+ \frac{0.05625}{3}(0.326212 + 4\, x 0.363267 + 0.412168)$$

$$+ \frac{0.05625}{3}(0.412168 + 4\, x 0.479399 + 0.577269)$$

$$= \frac{0.05625}{3}(1.534854 + 1.789377 + 2.191448 + 2.907033)$$

$$= \frac{0.05625}{3} x\, 8.422712 = 0.1579, \frac{kg\ catalyst}{m^2}$$

Hence

$$V = 10^5\, F_{Ao} x\, 0.1579 = 10^5\, x\, 1.6\, x 10^{-3} x\, 0.1579$$
$$= 25.264\, \frac{kg\ catalyst}{m^2}$$

Since the molecular weight of ethyl benzene is 106 and that of styrene is 104, to produce 20,000 kg per 24 hour day of styrene would require

$$\frac{106}{104} x \frac{20{,}000}{24\ x\ 3600} = 0.2359 \frac{kg}{s}\ ethyl\ benzene\ per\ day$$

$$= 0.2359 \frac{kg}{s} x \frac{kmol}{106\ kg} = 2.226\ x10^{-3}\ \frac{kmol}{s}\ ethyl\ benzene$$

Since the feed rate of ethyl benzene is $1.6\ x10^{-3}\ \frac{kmol}{m^2.s}$, total cross-sectional area of the tubes is

$$\frac{2.226\ x10^{-3}}{1.6\ x10^{-3}}, \frac{kmol}{s}. \frac{m^2.s}{kmol} = 1.391\ m^2$$

Since each tube is given to be 80 mm diameter, the number of tubes, n, is

$$n = \frac{1.391\ x\ 4}{\pi\ x\ (0.080)^2} = 276.7\ \cong 277\ tubes$$

The total volume of the reactor, V, will be, in terms of the quantity of catalyst it contains

$$25.264, \frac{kg\ catalyst}{m^2} x\ 1.391\ m^2 = 35.142\ kg\ catalyst$$

Since the bulk density of the catalyst is 1440 kg/m^3, the total volume of the reactor occupied by catalysts is

$$\frac{35.142}{1440}\ kg\ catalyst \frac{m^3}{kg\ catalyst} = 0.0244\ m^3$$

Thus, length of tubes, L, is given by

$$L = \frac{Volume\ x\ n}{Total\ Cross - sectional\ area} = \frac{0.0244\ x\ 276.7}{1.391}$$

$$= 4.854\ m\ Ans$$

Example 4.17

Ethyl formate is to be produced from ethanol and formic acid in a continuous tubular flow reactor operated at a constant temperature of 30 C. The reactants will be fed to the reactor in the proportions 1 mol of HCOOH to 5 mols C_2H_5OH at a combined flow rate of 0.72 m^3/h. The reaction will be catalysed by a small amount of sulphuric acid. At the temperature, mole ration and catalyst concentration to be employed, the rate equation, determined from small scale batch experiments, has been found to be

$$-r_A = kC_A^2$$

where $-r_A$= kmol of formic acid reacting per m^3 per second, C_A is the concentration of formic acid, $kmol/m^3$ and k = 2.8 x 10^{-4} $m^3/kmol.s$. The density of the mixture is 820 kg/m^3 and may be assumed constant throughout.

Estimate the volume of the reactor required to convert 70 % of the formic acid to the ester. If the reactor consists of a pipe of 50 mm ID, what will be the total length required? Determine, also, whether the flow will be laminar or turbulent and comment on the significance of your conclusion in relation to your estimate of reactor volume. The viscosity of the solution is 1.4 x 10^{-3} Ns/m^2. (Fogler, 1974)

Answer

The reaction equation may be expressed as

$$HCOOH + C_2H_5OH = C_2H_5.COOH + H_2O$$

$$Formic\ acid\ +\ Ethanol = Ethyl\ formate + Water$$

If we represent formic acid by A, then the volume of the plug flow reactor required is

$$V = \int_0^{X_A} \frac{F_{Ao} dX_A}{-r_A} \qquad (1)$$

The rate equation we were given

$$-r_A = kC_A^2$$

may be stated in terms of fractional conversion as

$$-r_A = kC_{Ao}^2(1 - X_A)^2 \qquad (2)$$

Substituting (2) in (1) and integrating

$$V = \frac{F_{Ao}}{kC_{Ao}^2}\int_0^{X_{Af}} \frac{dX_A}{(1 - X_A)^2} = \frac{F_{Ao}}{kC_{Ao}^2}\left|\frac{1}{1 - X_A}\right|_0^{X_{Af}}$$

$$V = \frac{F_{Ao}}{kC_{Ao}^2}\left(\frac{1}{1 - X_{Af}} - 1\right) = \frac{F_{Ao}}{kC_{Ao}^2}\frac{X_{Af}}{1 - X_{Af}} \qquad (3)$$

The combined mass flow rate, F_T, at a constant density of 820 kg/m^3, was given as 0.72 m^3/h and, in kg/h, is

$$F_T = 0.72\,x\,820, \frac{m^3}{h}.\frac{kg}{m^3} = 590.4\frac{kg}{h} \qquad (4)$$

If F_{Ao} is the molar flow rate of formic acid per hour, then 5 F_{Ao} is the molar flow rate per hour of ethanol.

Since the formula weight of formic acid is 46 kg and that of ethanol is 46 kg,

$$46F_{Ao} + 5\,x\,46\,F_{Ao} = 590.4$$

from which

$$F_{Ao} = \frac{590.4}{6\,x\,46} = 2.139\frac{kmol}{h} \qquad (5)$$

But

$$C_{Ao} = \frac{F_{Ao}}{0.72}, \frac{kmol}{h}.\frac{h}{m^3} = \frac{F_{Ao}}{0.72}\frac{kmol}{m^3} \qquad (6)$$

Substituting (5), (6) and the given values of k and X_{Af} in equation (3)

$$V = \frac{F_{Ao}}{kC_{Ao}^2}\frac{X_{Af}}{1 - X_{Af}} = \frac{F_{Ao}(0.72)^2}{kF_{Ao}^2 x 3600}\frac{X_{Af}}{1 - X_{Af}}, \frac{kmol}{h}.\frac{kmol.s}{m^3}.\frac{h}{s}.\frac{m^6}{(kmol)^2}$$

$$= \frac{(0.72)^2}{kF_{Ao}x3600}\frac{X_{Af}}{1-X_{Af}} = \frac{(0.72)^2}{2.8\,x\,10^{-4}x\,2.139\,x3600}\cdot\frac{0.7}{1-0.7}$$

$$= 0.561m^3 \quad Ans$$

If the pipe inner diameter, d, is 50 mm, then length of pipe, L, is given by

$$V = 0.561 = \frac{\pi d^2 L}{4}$$

from which

$$L == \frac{4\,x\,0.561}{\pi(0.050)^2} = 285.71\,m \quad Ans$$

To determine whether the flow was laminar or turbulent, we would need to estimate the Reynolds's, Re, number for the flow. Thus

$$Re = \frac{\rho du}{\mu} \tag{7}$$

since

$$F_V = 0.72 = \frac{\pi d^2 u}{4}$$

$$u = \frac{4\,x\,0.72}{\pi d^2} \tag{8}$$

Substituting (8) in (7) with the given numerical values

$$Re = \frac{\rho du}{\mu} = \frac{\rho\,x\,4\,x\,0.72}{\pi\,x\,\mu\,x\,d}$$

$$= \frac{820\,x\,4\,x\,0.72}{\pi\,x\,0.0014\,x\,0.050\,x\,3600}, \frac{kg}{m^3}\frac{m^3}{h}\frac{m.s}{kg}\frac{1}{m}\frac{h}{s} = 2983 \tag{9}$$

This value of the Reynolds's number lies between 2000 and 4000 which is the regime of transition flow from laminar to turbulent flow.

The volume of the plug flow reactor stated in equation (1) is based on the plug flow assumption. This assumption implies a flat velocity profile which, in fact, is perfect turbulent flow.

The reality, however, as shown by the value of the Reynolds's number, is that the flow is not turbulent. Hence the value of V calculated is under estimated.

Example 4.18

Two stirred tanks are available at a chemical works, one of volume 100 m^3, the other 30 m^3. It is suggested that these tanks be used as a two stage CSTR for carrying out an irreversible liquid phase reaction

$$A + B \rightarrow products$$

The two reactants will be present in the feed stream in equimolar proportion, the concentration of each being 1.5 kmol/m^3. The volumetric flow rate of the feed stream will be 0.3 x 10^{-3} m^3/s.

The reaction is irreversible and is first order with respect to each of the reactants (i.e second order overall) with a rate constant of 1.8 x 10^{-4} m^3/kmol.s.

a. Which tank should be used as the first stage of the reactor system in order to get the highest conversion possible?
b. With this configuration, calculate the conversion obtained in the product stream leaving the second tank after steady state conditions have been reached. (Fogler, 1974)

Answer

For the given reaction

$$A + B \rightarrow products$$

$$-r_A = k[A][B] = kC_{Ao}^2(1 - X_A)(\theta_B - X_A) \qquad (1)$$

where k = 1.8 x 10^{-4} m^3/kmol.s

Case a

For the reaction system shown below

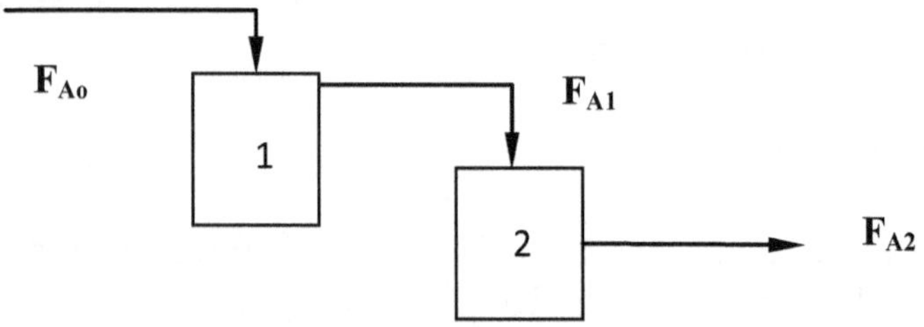

The reaction rate – conversion plot could be as shown below, with another possibility being CSTR2 having the shape of CSTR1 and vice versa.

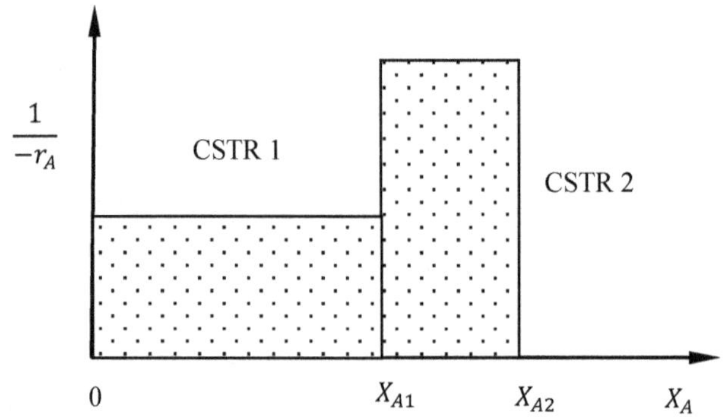

Suppose the first tank is 100 m^3 and the second tank, 30 m^3. Since

$$V_1 = \frac{F_{Ao}X_{A1}}{-r_{A1}} \tag{2}$$

$$F_{Ao} = u_o C_{Ao} = 3 \; x \; 10^{-4} \; x \; 1.5 \, , \frac{m^3}{s}\frac{kmol}{m^3} = 4.5 \; x \; 10^{-4} \frac{kmol}{s} \tag{3}$$

Since $\theta_B = 1$

$$-r_{A1} = kC_{Ao}^2(1 - X_{A1})^2$$

$$= 1.8 \; x \; 10^{-4} \; x \; (1.5)^2(1 - X_{A1})^2 \, , \frac{m^3}{kmol.s}\left(\frac{kmol}{m^3}\right)^2$$

$$= 4.05\ x\ 10^{-4}(1 - X_{A1})^2\ \frac{kmol}{m^3.s} \qquad (4)$$

Substituting (3) and (4) and the given values in (2)

$$100 = \frac{4.5\ x\ 10^{-4} X_{A1}}{4.05\ x\ 10^{-4}(1 - X_{A1})^2}, \frac{kmol}{s} \frac{m^3.s}{kmol}$$

from which

$$\frac{X_{A1}}{(1 - X_{A1})^2} = 90$$

This results in the quadratic equation

$$90X_{A1}^2 - 181X_{A1} + 90 = 0$$

which has the solution

$$X_{A1} = \frac{181 \pm \sqrt{32761 - 32400}}{180} = \frac{181 \pm 19}{180} = 1.11\ \ or\ \ 0.9$$

Since mole fraction cannot be greater than 1, $X_{A1} = 0.9$

For the second tank

$$V_2 = \frac{F_{A1} X_{A2}}{-r_{A2}} \qquad (5)$$

$$F_{A1} = F_{Ao}(1 - X_{A1}) = 4.5\ x\ 10^{-4}\ x\ 0.1 = 0.45\ x\ 10^{-4}\frac{kmol}{s} \qquad (6)$$

$$C_{A1} = \frac{F_{A1}}{u_o} = \frac{0.45\ x\ 10^{-4}}{3\ x\ 10^{-4}}, \frac{kmol}{s} \frac{s}{m^3} = 0.15\ \ \frac{kmol}{m^3} \qquad (7)$$

$$-r_{A2} = kC_{A1}^2(1 - X_{A2})^2$$
$$= 1.8\ x\ 10^{-4}\ x\ (0.15)^2(1 - X_{A2})^2, \frac{m^3}{kmol.s}\left(\frac{kmol}{m^3}\right)^2$$

$$= 4.05\ x\ 10^{-6}(1 - X_{A2})^2\ \frac{kmol}{m^3.s} \qquad (8)$$

Substituting (6) and (8) and the given values in (5)

$$30 = \frac{0.45\ x\ 10^{-4} X_{A2}}{4.05\ x\ 10^{-6}(1 - X_{A2})^2}, \frac{kmol}{s} \frac{m^3.s}{kmol}$$

from which

$$\frac{X_{A2}}{(1 - X_{A2})^2} = 2.7$$

This results in the quadratic equation

$$2.7X_{A2}^2 - 6.4X_{A2} + 2.7 = 0$$

which has the solution

$$X_{A2} = \frac{6.4 \pm \sqrt{40.96 - 29.16}}{5.4} = \frac{6.4 \pm 3.44}{5.4} = 1.82\ or\ 0.55$$

Since mole fraction cannot be greater than 1, $X_{A2} = 0.55$

Thus

$$F_{A2} = F_{A1}(1 - X_{A2}) = F_{Ao}(1 - X_{A1})(1 - X_{A2})$$

Since u_o is constant in the system

$$Overall\ conversion = \frac{C_{Ao} - C_{A2}}{C_{Ao}} = \frac{F_{Ao} - F_{A2}}{F_{Ao}}$$

$$= \frac{F_{Ao} - F_{Ao}(1 - X_{A1})(1 - X_{A2})}{F_{Ao}} = 1 - (1 - X_{A1})(1 - X_{A2})$$

$$= 1 - (1 - 0.9)(1 - 0.55) = 0.955$$

If the first tank is 30 m^3 and the second tank 100 m^3, the calculations are as before.

Since

$$V_1 = \frac{F_{Ao}X_{A1}}{-r_{A1}} \tag{9}$$

$$F_{Ao} = u_o C_{Ao} = 3\,x\,10^{-4}\,x\,1.5\,, \frac{m^3}{s}\frac{kmol}{m^3} = 4.5\,x\,10^{-4}\frac{kmol}{s} \qquad (10)$$

Since $\theta_B = 1$

$$-r_{A1} = kC_{Ao}^2(1 - X_{A1})^2$$

$$= 1.8\,x\,10^{-4}\,x\,(1.5)^2(1 - X_{A1})^2\,, \frac{m^3}{kmol.s}\left(\frac{kmol}{m^3}\right)^2$$

$$= 4.05\,x\,10^{-4}(1 - X_{A1})^2\,\frac{kmol}{m^3.s} \qquad (11)$$

Substituting (10) and (11) and the given values in (9)

$$30 = \frac{4.5\,x\,10^{-4}X_{A1}}{4.05\,x\,10^{-4}(1 - X_{A1})^2}, \frac{kmol}{s}\frac{m^3.s}{kmol}$$

from which

$$\frac{X_{A1}}{(1 - X_{A1})^2} = 27$$

This results in the quadratic equation

$$27X_{A1}^2 - 55X_{A1} + 27 = 0$$

which has the solution

$$X_{A1} = \frac{55 \pm \sqrt{3025 - 2916}}{54} = \frac{55 \pm 10.44}{54} = 1.21\ \ or\ \ 0.83$$

Since mole fraction cannot be greater than 1, $X_{A1} = 0.83$

For the second tank

$$V_2 = \frac{F_{A1}X_{A2}}{-r_{A2}} \qquad (12)$$

$$F_{A1} = F_{Ao}(1 - X_{A1}) = 4.5\,x\,10^{-4}\,x\,0.17 = 0.765\,x\,10^{-4}\frac{kmol}{s} \qquad (13)$$

$$C_{A1} = \frac{F_{A1}}{u_o} = \frac{0.765\,x\,10^{-4}}{3\,x\,10^{-4}}, \frac{kmol}{s}\frac{s}{m^3} = 0.255\ \frac{kmol}{m^3} \qquad (14)$$

$$-r_{A2} = kC_{A1}^2(1 - X_{A2})^2$$

$$= 1.8\,x\,10^{-4}\,x\,(0.255)^2(1 - X_{A2})^2\,,\frac{m^3}{kmol.s}\left(\frac{kmol}{m^3}\right)^2$$

$$= 1.17\,x\,10^{-5}(1 - X_{A2})^2\,\frac{kmol}{m^3.s} \qquad (15)$$

Substituting (13) and (15) and the given values in (12)

$$100 = \frac{0.765\,x\,10^{-4}X_{A2}}{1.17\,x\,10^{-5}(1 - X_{A2})^2},\frac{kmol}{s}\,\frac{m^3.s}{kmol}$$

from which

$$\frac{X_{A2}}{(1 - X_{A2})^2} = 15.29$$

This results in the quadratic equation

$$15.29X_{A2}^2 - 31.59X_{A2} + 15.29 = 0$$

which has the solution

$$X_{A1} = \frac{31.59 \pm \sqrt{997.928 - 935.136}}{30.58} = \frac{31.59 \pm 7.92}{30.58}$$
$$= 1.29\ or\ 0.77$$

Since mole fraction cannot be greater than 1, $X_{A2} = 0.77$

Thus

$$F_{A2} = F_{A1}(1 - X_{A2}) = F_{Ao}(1 - X_{A1})(1 - X_{A2})$$

Since u_o is constant in the system

$$Overall\ conversion = \frac{C_{Ao} - C_{A2}}{C_{Ao}} = \frac{F_{Ao} - F_{A2}}{F_{Ao}}$$

$$= \frac{F_{Ao} - F_{Ao}(1 - X_{A1})(1 - X_{A2})}{F_{Ao}} = 1 - (1 - X_{A1})(1 - X_{A2})$$

$$= 1 - (1 - 0.83)(1 - 0.77) = 0.961$$

Thus using the 30 m^3 tank as the first tank gives the highest possible conversion.

Case b

$$Conversion\ in\ last\ tank = \frac{C_{A1} - C_{A2}}{C_{A1}} = \frac{F_{A1} - F_{A2}}{F_{A1}}$$

$$= \frac{F_{A1} - F_{A1}(1 - X_{A2})}{F_{A1}} = 1 - (1 - X_{A2})$$

$$= 1 - (1 - 0.77) = 0.77 \quad Ans$$

References for Chapter Four

1. B. N. Nnolim: Unpublished Lecture Notes in Chemical Reaction Engineering; IMT, Enugu, Nigeria; 1989
2. Denbigh K. G. & Turner J. C. R.; *Chemical Reactor Theory*; 2nd edition, London; Cambridge University Press, 1971
3. Fogler, H Scott, *The Elements of Chemical Kinetics and Reactor Calculations*; Prentice Hall Inc; N. J., USA, 1974
4. Levenspiel O; *Chemical Reaction Engineering*; Wiley International Edition; New York, USA, 1972
5. Walas S. M.; *Reaction Kinetics for Chemical Engineers*; McGraw-Hill Book Company, New York, USA, 1959

CHAPTER FIVE:
ENERGY BALANCES IN CHEMICAL REACTORS

Example 5.1

The heat of reaction at the temperature, T, for any reaction, may be found from the equation

$$\Delta H_R^T = \sum_{i=1}^{n} v_i\left(\Delta H_i^f\right)_{298} + \int_{298}^{T} \Delta Cp dT$$

where

ΔCp	$\sum_{i=1}^{n} v_i Cp_i$
v_i	Stoichiometric coefficient of component i
Cp_i	heat capacity per mole of component i
$\left(\Delta H_i^f\right)_{298}$	Heat of formation of component i at 298 K
ΔH_R^T	Heat of reaction per mole at temperature T

For the reaction

$$CH_4(g) + H_2O(g) \leftrightarrow CO(g) + 3H_2(g)$$

Calculate the heat of reaction, at 298 K, per mole of methane given that the heats of formation, at 298 K, are as given below.

Component	$\left(\Delta H_i^f\right)_{298}, kJ/kmol$
$CO(g)$	*-110,621*
$H_2(g)$	*0.0*
$CH_4(g)$	*-74,901*
$H_2O(g)$	*-242,009*

Answer

Since T = 298 K, $\int_{298}^{T} \Delta Cp dT = 0$. Hence

$$\Delta H_R^T = \sum_{i=1}^{n} v_i (\Delta H_i^f)_{298}$$

For the given reaction

$$\Delta H_R^{298} = \Delta H_{CO}^{f,298} + 3\Delta H_{H_2}^{f,298} - \Delta H_{H_2O}^{f,298} - \Delta H_{CH_4}^{f,298}$$

$$= -110,621 + 3 \, x \, 0 - (-242,009) - (-74,901)$$

$$= 206,289 \frac{kJ}{kmol} \quad Ans$$

Example 5.2

The combustion of carbon monoxide in oxygen is described as a chemical reaction. Thus

$$CO + ½O_2 = CO_2$$

What would be the heat of reaction and combustion at 673 K given that the heats of formation, ΔH_f^{298}, and heat capacities are as follows

	CO	O_2	CO_2
ΔH_f^{298}, J/mol	-110,621	0.0	-393,796
Cp, J/mol	$26.53 + 7.70 \, x \, 10^{-3}T - 1.17 \, x \, 10^{-6}T^2$	$25.52 + 13.60 \, x \, 10^{-3}T - 4.27 \, x \, 10^{-6}T^2$	$26.78 + 42.26 \, x \, 10^{-3}T - 14.23 \, x \, 10^{-6}T^2$

Answer

By the Kirchoff's equation

$$\Delta H_R^T = \Delta H_R^{298} + \int_{298}^{T} \Delta Cp\, dT \qquad (1)$$

$$\Delta H_R^{298} = \Delta H_{f,CO_2}^{298} - \Delta H_{f,CO}^{298} - ½\Delta H_{f,O_2}^{298}$$

$$= -393,796 - (-110,621) - ½\ x\ 0.0 = -283,175 \frac{J}{mol} \qquad (2)$$

$$\Delta Cp = (Cp)_{CO_2} - (Cp)_{CO} - ½(Cp)_{O_2}$$

$$= 26.78 + 42.26\ x\ 10^{-3}T - 14.23\ x\ 10^{-6}T^2$$

$$-26.53 - 7.70\ x\ 10^{-3}T + 1.17\ x\ 10^{-6}T^2$$

$$-½(25.52 + 13.60\ x\ 10^{-3}T - 4.27\ x\ 10^{-6}T^2)$$

$$= -12.51 + 27.76\ x\ 10^{-3}T - 10.925\ x\ 10^{-6}T^2, \ \frac{J}{mol.K} \qquad (3)$$

$$\int_{298}^{673} \Delta Cp\, dT = -12.51(673 - 298) + 27.76\ x\ 10^{-3}\left(\frac{673^2}{2} - \frac{298^2}{2}\right)$$

$$- 10.925\ x\ 10^{-6}\left(\frac{673^3}{3} - \frac{298^3}{3}\right)$$

$$= -4,691.25 + 5,054.055 - 1,013.686$$

$$= -650.881, \ \frac{J}{mol} \qquad (4)$$

Substituting (2) and (4) in (1)

$$\Delta H_R^{673} = -283,175 - 650.881 = -283,826, \frac{J}{mol} \qquad Ans$$

Example 5.3

Show that for the generalised reaction

$$aA + bB \rightarrow cC + dD$$

taking place in a steady flow reactor,

$$Q = \sum_{i=1}^{n} F_{io}(H_i - H_{io}) + F_{Ao}X_A\Delta H_R^T$$

where

Q	Total rate of heat addition or removal
F_{io}	Molar flow rate of specie, i, into the reactor
F_{Ao}	Molar flow rate of key reactant, A, into the reactor
H_i , H_{io}	Molal enthalpies of specie i at exit and entrance, respectively, to the reactor
X_A	Fractional conversion of key reactant, A
ΔH_R^T	Heat of reaction at temperature, T, per mole of key component reacted

Are there any simplifications you can make?

Answer

The steady state energy balance for a flow system is given, from the first law of thermodynamics, by

$$\begin{aligned} Total\ Heat\ absorbed - Total\ Work\ done \\ = Sum\ of\ all\ energy\ entering\ the\ system \\ - Sum\ of\ all\ energy\ leaving\ the\ system \end{aligned}$$

That is

$$\sum Q - \sum W_S = \sum \left(H + gz + \frac{v^2}{2}\right)_{out} F_{out} - \sum \left(H + gz + \frac{v^2}{2}\right)_{in} F_{in}$$

Since, in this problem, $W_S = 0; gz = 0; \frac{v^2}{2} = 0$

$$\sum Q = \sum H_{out} F_{out} - \sum H_{in} F_{in} \qquad (1)$$

We can rewrite the reaction equation as

$$A + \frac{b}{a}B \rightarrow \frac{c}{a}C + \frac{d}{a}D$$

so that, at conversion X_A,

$$F_A = F_{Ao}(1 - X_A);\; F_B = F_{Bo} - \frac{b}{a} F_{Ao}X_A;\; F_C = F_{Co} + \frac{c}{a} F_{Ao}X_A;$$
$$F_D = F_{Do} + \frac{d}{a} F_{Ao}X_A;\; F_I = F_{Io} \qquad (2)$$

where the second subscript, o, indicates an initial value and no second subscript indicates a final or exit value.

The elaboration of the energy balance of equation (1) is

$$Q + H_{Ao}F_{Ao} + H_{Bo}F_{Bo} + H_{Co}F_{Co} + H_{Do}F_{Do} + H_{Io}F_{Io}$$
$$= H_AF_A + H_BF_B + H_CF_C + H_DF_D + H_IF_I \qquad (3)$$

where the H_{Ao}, H_{Bo}, H_{Co}, H_{Do} and H_{Io} are initial specific enthalpy values and H_A, H_B, H_C, H_D and H_I are exit specific enthalpy values.

When we substitute the values from equation (2) into equation (3), we get

$$Q + H_{Ao}F_{Ao} + H_{Bo}F_{Bo} + H_{Co}F_{Co} + H_{Do}F_{Do} + H_{Io}F_{Io}$$
$$= H_AF_{Ao}(1 - X_A) + H_B\left(F_{Bo} - \frac{b}{a}F_{Ao}X_A\right) + H_C\left(F_{Co} + \frac{c}{a}F_{Ao}X_A\right)$$
$$+ H_D\left(F_{Do} + \frac{d}{a}F_{Ao}X_A\right) + H_IF_I$$

which give, on simplification

$$Q + H_{Ao}F_{Ao} + H_{Bo}F_{Bo} + H_{Co}F_{Co} + H_{Do}F_{Do} + H_{Io}F_{Io}$$
$$= H_AF_{Ao} + H_BF_{Bo} + H_CF_{Co} + H_DF_{Do} + H_IF_I$$
$$-F_{Ao}X_A\left(H_A + \frac{b}{a}H_B - \frac{c}{a}H_C - \frac{d}{a}H_D\right)$$

Thus

$$Q = F_{Ao}(H_A - H_{Ao}) + F_{Bo}(H_B - H_{Bo}) + F_{Co}(H_C - H_{Co})$$
$$+F_{Do}(H_D - H_{Do}) + F_{Io}(H_I - H_{Io})$$
$$+F_{Ao}X_A\left(\frac{c}{a}H_C + \frac{d}{a}H_D - \frac{b}{a}H_B - H_A\right) \qquad (4)$$

But the heat of reaction at the temperature, T, is given by

$$\Delta H_R^T = \frac{c}{a}H_C + \frac{d}{a}H_D - \frac{b}{a}H_B - H_A$$

Equation (4) reduces, therefore, to

$$Q = \sum_{i=1}^{n} F_{io}(H_i - H_{io}) + F_{Ao}X_A\Delta H_R^T \qquad (5)$$

where i represents the species A, B, C, D and I. Ans

<u>Simplifications</u>

<u>If there is no change in phase</u>

$$H_i(T) = H_i^o(T_R) + \int_{T_R}^{T} Cp_i\, dT$$

Then

$$H_i(T) - H_{io}(T_o) = H_i^o(T_R) + \int_{T_R}^{T} Cp_i\, dT - H_i^o(T_R) - \int_{T_R}^{T_o} Cp_i\, dT$$

$$= \int_{T_o}^{T} Cp_i\, dT \qquad (6)$$

Substituting (6) in (5)

$$Q = \sum_{i=1}^{n} F_{io}\int_{T_o}^{T} Cp_i\, dT + F_{Ao}X_A\Delta H_R^T \qquad (7)$$

Since

$$\Delta H_R^T = \Delta H_i^o(T_R) + \int_{T_R}^{T} \Delta Cp_i\, dT$$

and

$$Cp_i = \alpha_i + \beta_i T + \gamma_i T^2$$

so that

$$mean\ Cp = \overline{\Delta Cp} = \frac{\int_{T_R}^{T} \Delta Cp\, dT}{\int_{T_R}^{T} dT} \tag{8}$$

$$\Delta H_R^T = \Delta H_i^o(T_R) + \overline{\Delta Cp}\ (T - T_R) \tag{9}$$

Also

$$\int_{T_o}^{T} Cp_i\, dT = \overline{Cp_i}\ x\ \Delta T_i \tag{10}$$

If we define

$$\Delta T_i = T - T_{io} \quad and\ \theta_i = \frac{F_i}{F_{Ao}}$$

equation (7) becomes

$$Q = \sum_{i=1}^{n} F_{io}\overline{Cp_i}\,.\,\Delta T_i + F_{Ao}X_A\ [\Delta H_i^o(T_R) + \overline{\Delta Cp}\ (T - T_R)]$$

$$\frac{Q}{F_{Ao}} = \sum_{i=1}^{n} \frac{F_{io}}{F_{Ao}}\overline{Cp_i}\,.\,\Delta T_i + X_A\ [\Delta H_i^o(T_R) + \overline{\Delta Cp}(T - T_R)\ ]$$

$$\frac{Q}{F_{Ao}} - X_A\ [\Delta H_i^o(T_R) + \overline{\Delta Cp}(T - T_R)\ ] = \sum_{i=1}^{n} \theta_i\overline{Cp_i}\,.\,\Delta T_i \qquad Ans$$

Example 5.4

The steady state energy balance equation around a chemical reactor, when kinetic, potential and viscous dissipation energies are negligible, is given by

$$\frac{Q}{F_{Ao}} - \frac{W_S}{F_{Ao}} - X_A\ [\Delta H_i^o(T_R) + \overline{\Delta Cp}(T - T_R)\ ] = \sum_{i=1}^{n} \theta_i\overline{Cp_i}\,.\,\Delta T_i$$

where

Q	rate of heat release or absorption
F_{Ao}	molar flow rate of key reactant, A, into the reactor
W_S	shaft work done

X_A	fractional conversion of key reactant, A
$\Delta H_i^o(T_R)$	heat of reaction at temperature, T_R, per mole of key component reacted
$\overline{\Delta Cp}$	mean heat capacity difference between T_R and T
θ_i	ratio of molar flow rate of specie i to that of the key component
$\overline{Cp_i}$	mean heat capacity of specie i between inlet and reaction temperatures
ΔT_i	temperature difference between reaction temperature and inlet temperature of specie i

Determine the steady state reaction temperature for the elementary, liquid phase reaction

$$A + B = 2C$$

taking place in a CSTR, jacketed with steam. You are given that

Reactor volume	$0.475\ m^3$
Steam jacket area	$0.929\ m^2$
Jacket steam saturation temperature	458.5 K
Overall heat transfer coefficient of jacket	$U = 852\ W/m^2K$
Agitator shaft horsepower	25 HP
Heat of reaction (independent of temperature)	+46520 kJ/kmol of A

You are, also, given the characteristics of the reactants and products as

	A	B	C
Feed, kmol/h	4.55	4.55	0
Feed temperature, K	300	300	
Specific heat, kJ/kmol.K	213.5	184.2	198.9
Molecular weight, kg	128	94	
Density, kg/m^3	1009	1077	1041

Take 1 HP = 745 Watts

Answer

$$\dot{Q} = -U\,A\,(T - T_C) = -\frac{852}{1000}\,x\,0.929(T - 458.5)\;kW$$

$$= -0.792(T - 458.5)\;kW \qquad (1)$$

$$W_S = 25\,x\,0.745 = 18.625\;kW \qquad (2)$$

$$\overline{\Delta Cp} = 2Cp_C - Cp_A - Cp_B = 2\,x\,198.9 - 213.5 - 184.2 = 0 \qquad (3)$$

$$\Delta H_i^o(T_R) + \overline{\Delta Cp}(T - T_R) = 46520\frac{kJ}{kmol} \qquad (4)$$

$$\theta_i = \frac{F_i}{F_{Ao}} = \frac{4.55}{4.55} = 1 \qquad (5)$$

$$\sum_{i=1}^{n} \theta_i \overline{Cp_i}\,.\Delta T_i = (0 + 213.5 + 184.2)(T - 300)$$

$$= 397.7(T - 300)\;\frac{kJ}{kmol} \qquad (6)$$

$$F_{Ao} = \frac{4.55}{3600} = 1.264\,x\,10^{-3}\;\frac{kmol}{s} \qquad (7)$$

Substituting (1) to (7) in the given equation

$$\frac{-0.792(T - 458.5)}{1.264\,x\,10^{-3}} - \frac{18.625}{1.264\,x\,10^{-3}} - X_A[46520] = 397.7(T - 300)$$

That is

$$-626.58(T - 458.5) - 14734.97 - 46520X_A = 397.7(T - 300)$$

As this is not a reversible reaction, a steady temperature will occur when $X_A = 1$. Thus

$$-626.58T + 287286.93 - 14734.97 - 46520 = 397.7T - 119310$$

That is

$$1024.28T = 345341.96$$

or

$$T = 337.16\,K \quad Ans$$

Example 5.5

The reversible first order, exothermic, reaction

$$A \underset{k_2}{\overset{k_1}{\rightleftharpoons}} B$$

is carried out, adiabatically, in a CSTR. The feed is at 294 K and is a 1.26 kg/s of a solution containing only A at a concentration of 1.6 kmol/m^3. Find the volume of the reactor necessary for 90 % conversion of A. You are given the following data

Solution density, kg/m^3	897
Heat capacity of solution, kJ/kg.K	2.931
Forward reaction rate constant, k_1, at 298 K, per hour	1.2
Activation energy, kJ/kmol	104, 675
Heat of Reaction, ΔH_R^{298}, kJ/kmol	-18, 608
Equilibrium constant, K_E	12.2
Gas constant, R, kJ/kmol.K	8.314

Answer

For a CSTR

$$V = \frac{F_{Ao}X_A}{-r_A} \tag{1}$$

The only unknown in equation (1) is $-r_A$.

<u>To determine</u> $-r_A$

For the reaction given, the rate equation may be expressed as

$$-r_A = k_1C_A - k_2C_B = k_1C_{Ao}(1 - X_A) - k_2C_{Bo}(1 - X_B) \tag{2}$$

Since -d$C_A = dC_B$

$$C_{Ao}X_A = -C_{Bo}X_B \tag{3}$$

That is

$$X_B = -\frac{C_{Ao}X_A}{C_{Bo}} \tag{4}$$

Putting (4) in (2)

$$-r_A = k_1C_{Ao}(1 - X_A) - k_2C_{Bo}\left(1 + \frac{C_{Ao}X_A}{C_{Bo}}\right)$$

$$= k_1C_{Ao}(1 - X_A) - k_2C_{Bo}\left(\frac{C_{Bo}+C_{Ao}X_A}{C_{Bo}}\right)$$

$$= k_1C_{Ao}(1 - X_A) - k_2C_{Ao}(\theta + X_A) \tag{5}$$

where

$$\theta = \frac{C_{Bo}}{C_{Ao}} \tag{6}$$

At any $-r_A$ and temperature, T

$$\frac{k_1}{k_2} = K_T \qquad (7$$

At equilibrium, $-r_A = 0,\ X_A = X_{AE},\ K_T = K_E$. Hence, from (5)

$$\frac{k_1}{k_2} = K_E = \frac{\theta + X_{AE}}{1 - X_{AE}} \tag{8}$$

from which

$$X_{AE} = \frac{K_E - \theta}{1 + K_E} \tag{9}$$

Also, from (9) and (7)

$$\theta + X_A = \frac{k_1}{k_2} - X_{AE}\left(1 + \frac{k_1}{k_2}\right) + X_A = \frac{k_1 - k_2X_{AE} - k_1X_{AE} + k_2X_A}{k_2}$$

$$= \frac{k_1(1 - X_{AE}) - k_2X_{AE} + k_2X_A}{k_2} \tag{10}$$

Putting (10) in (5)

$$-r_A = k_1C_{Ao}(1 - X_A) - k_2C_{Ao}\left[\frac{k_1(1 - X_{AE}) - k_2X_{AE} + k_2X_A}{k_2}\right]$$

$$= k_1C_{Ao}(1 - X_A) - C_{Ao}k_1(1 - X_{AE}) + k_2C_{Ao}X_{AE} - k_2C_{Ao}X_A$$

$$= k_1 C_{Ao}(X_{AE} - X_A) + k_2 C_{Ao}(X_{AE} - X_A)$$

That is

$$-r_A = (k_1 + k_2) C_{Ao}(X_{AE} - X_A) \qquad (11)$$

To determine k_1 and k_2

Since

$$k_1 = A_1 e^{-E_1/RT} \qquad (12)$$

At 298 K, with the given values,

$$\frac{1.2}{h} \cdot \frac{h}{3600s} = 3.333\, x\, 10^{-4} = A_1 e^{-104675/8.314\, x\, 298} = A_1 e^{-42.249}$$

$$A_1 = 7.436\, x\, 10^{14}\; per\ second \qquad (13)$$

For any T, therefore, equation (12) becomes

$$k_1 = 7.436\, x\, 10^{14} e^{-12{,}590/T} \qquad (14)$$

To determine the temperature, T, of the adiabatic reaction

An energy balance for the CSTR is

$$\sum F_i Cp_i \Delta T_i = -X[\Delta H_R^T + \Delta Cp(T - T_R)] F_{Ao} \qquad (15)$$

where F, Cp, ΔT, X, are flow rates, heat capacities, sensible temperature differences and fractional conversion, subscript i denotes component i, T_R is reaction temperature and T the adiabatic temperature reached in the reactor.

Since we are given that ΔH_R^T is constant, $\Delta Cp = 0$

$$F_{Ao} = u_o C_{Ao} = \frac{1.26}{897}\, x\, 1.60, \frac{kg}{s}\frac{m^3}{kg}\frac{kmol}{m^3} = 2.25\, x\, 10^{-3}\frac{kmol}{s} \qquad (16)$$

$$\sum F_i Cp_i \Delta T_i = 1.26\, x\, 2.931\, x\, (T - 294), \frac{kg}{s}\frac{kJ}{kg.K}.K$$

$$= 3.693(T - 294), \frac{kJ}{s} \quad (17)$$

Putting (16) and (17) in (15)

$$3.693(T - 294), \frac{kJ}{s} = -0.9\, x - 18,608\, x\, 2.25\, x\, 10^{-3}, \frac{kJ}{kmol} \cdot \frac{kmol}{s}$$

from which

$$T - 294 = 10.20$$

That is

$$T = 304.20\, K \quad (18)$$

Putting (18) in (14)

$$k_1 = 7.436\, x\, 10^{14} e^{-12,590/304.20} = 7.890\, x\, 10^{-4} \quad s^{-1} \quad (19)$$

From the van't Hoff equation

$$\frac{dln\, K}{dT} = \frac{\Delta H_R^T}{RT^2}$$

That is

$$\ln \frac{K_2}{K_1} = -\frac{\Delta H_R}{R}\left(\frac{1}{T_2} - \frac{1}{T_1}\right) \quad (20)$$

At 304.20 K

$$\ln \frac{K_2}{K_1} = -\frac{-18,608}{8.314}\left(\frac{1}{304.2} - \frac{1}{298}\right) = -0.1531$$

$$\frac{K_2}{K_1} = 0.8581 \quad (21)$$

Thus, since $K_{298} = K_1 = 12\ 2$, given

$$K_2 = 0.8581\, x\, K_1 = 0.8581\, x\, 12.2 = 10.468 \quad (22)$$

From (9) and (22) since $\theta = 0$ and $K_2 = K_E$ at equilibrium,

$$X_{AE} = \frac{K_E - \theta}{1 + K_E} = \frac{10.468}{1 + 10.468} = 0.913 \quad (23)$$

From (7)

$$\frac{k_1}{k_2} = K_2 \quad or \quad k_2 = \frac{7.890\ x\ 10^{-4}\ s^{-1}}{10.468} = 7.537\ x\ 10^{-5} s^{-1} \qquad (24)$$

Putting (19), (23) and (24) in (11),

$$-r_A = (k_1 + k_2)C_{Ao}(X_{AE} - X_A)$$

$$= (7.890\ x\ 10^{-4} + 7.537\ x\ 10^{-5})\ x\ 1.60\ x\ (0.913 - 0.9)$$

$$= 1.798\ x\ 10^{-5} s^{-1} \frac{kmol}{m^3.s} \qquad (25)$$

From (16), (25) and (1)

$$V = \frac{F_{Ao}X_A}{-r_A} = \frac{2.25\ x\ 10^{-3}\ x\ 0.9}{1.798\ x\ 10^{-5}} \frac{kmol}{s} \frac{m^3.s}{kmol} = 112.6\ m^3\ \ Ans$$

Example 5.6

The elementary gas reaction

$$A \underset{k_2}{\overset{k_1}{\rightleftarrows}} 2B$$

is to be carried out in the various reactors mentioned below. The feed, which is at a temperature of 300 K, consists of 30 mole % of A and the remainder inerts. The volumetric flow rate entering the reactor at this temperature is 0.472 m^3/s. The concentration of A in the feed at 300 K is 8.01 $kmol/m^3$. For 40 % conversion

a. Determine the volume of the plug flow reactor when the reaction is carried out adiabatically
b. Determine the volume of the CSTR which is operated adiabatically.

Take Cp_A = 58.15 kJ/kmol.K; Cp_B = 46.52 kJ/kmol.K; Cp_{Inerts} =

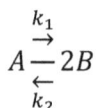

69.78 kJ/kmol.K.

The heat of reaction is a function of temperature and its value at 300 K is 14, 890 kJ/kmol of A. At 300 K, $k_1 = 0.217\ min^{-1}$; $K_C = 160.19\ kmol/m^3$. k_1 varies with temperature as follows.

T, K	300	340
k_1	0.217	0.324

Answer

For a steady state plug flow reactor

$$V = F_{Ao} \int_{X_{A1}}^{X_{A2}} \frac{dX_A}{-r_A} \quad (1)$$

<u>To determine r_A</u>

Because of the increase in the number of moles in the right hand side of the reaction equation, the change of volume with conversion has to be determined first.

Thus, a stoichiometric table is as follows

Basis: 100 moles of initial reaction Mixture

Species	**Moles initially present**	**Moles reacted**	**Moles remaining**
A	30	$-30X_A$	$30(1-X_A)$
B	0	$60X_A$	$60X_A$
Inerts	70	-	70
Total	100		$100 + 30X_A$

The volume of the reaction mixture is

$$V = V_o(1 + \epsilon X_A) \quad (2)$$

where

$$\epsilon = \delta y_{Ao} \quad (3)$$

$$\delta = \frac{2-1}{1} = 1 \tag{4}$$

Since $y_{Ao} = 0.3$, given

$$\epsilon = \delta y_{Ao} = 1\,x\,0.3 = 0.3 \tag{5}$$

Thus, from (2)

$$V = V_o(1 + \epsilon X_A) = V_o(1 + 0.3X_A) \tag{6}$$

The concentrations of reactants and products at any time are, therefore,

Species	Concentration, kmol/m^3
A	$\frac{30(1-X_A)}{V_o(1+0.3X_A)}$
B	$\frac{60X_A}{V_o(1+0.3X_A)}$
Inerts	$\frac{70}{V_o(1+0.3X_A)}$

Since the concentration of A in the feed at 300 K is given as 8.01 kmol/m^3,

$$C_{Ao} = \frac{30}{V_o} = 8.01\frac{kmol}{m^3} \quad or \quad V_o = 3.745\,m^3 \tag{7}$$

$$C_{Io} = \frac{70}{V_o} = \frac{70}{3.745} = 18.69\frac{kmol}{m^3} \tag{8}$$

For the reaction given, rate equation may be expressed as

$$-r_A = k_1C_A - k_2C_B^2 = k_1\frac{8.01(1-X_A)}{(1+0.3X_A)} - k_2\left(\frac{16.02X_A}{(1+0.3X_A)}\right)^2 \tag{9}$$

At equilibrium which can occur at any r_A and temperature, T

$$\frac{k_1}{k_2} = K_C \tag{10}$$

To determine k_1 as a function of temperature

Since

$$k_1 = A_1 e^{-E_1/RT} \tag{11}$$

At 300 K, with the given values,

$$\frac{0.217}{min} = A_1 e^{-E_1/8.314\,x\,300} = A_1 e^{-E_1/2494.2} \tag{11a}$$

At 340 K, with the given values,

$$\frac{0.324}{min} = A_1 e^{-E_1/8.314\,x\,340} = A_1 e^{-E_1/2826.8} \tag{11b}$$

Dividing (11b) by (11a)

$$\frac{0.324}{0.217} = 1.4931 = \frac{e^{-E_1/2826.8}}{e^{-E_1/2494.2}} = \exp(4.7173\,x\,10^{-5} E_1)$$

Thus

$$E_1 = 8{,}498.5\ kJ/kmol$$

and from (11a)

$$A_1 = \frac{0.217}{e^{-8498.5/2494.2}} = 6.55$$

For any T, therefore, equation (11) becomes

$$k_1 = 6.55 e^{-1022/T}, min^{-1} \tag{12}$$

To determine the temperature, T, of the adiabatic reaction

An energy balance for the CSTR is

$$\sum F_i Cp_i \Delta T_i = -X[\Delta H_R^T + \Delta Cp(T - T_R)]F_{Ao} \tag{13}$$

where F, Cp, ΔT, X, are flow rates, heat capacities, sensible temperature differences and fractional conversion, subscript i denotes component i, T_R is reaction temperature and T the adiabatic temperature reached in the reactor.

Since we are given that ΔH_R^T varies with temperature,

$$\Delta Cp = 2Cp_B + Cp_{Inert2} - Cp_A - Cp_{Inert1}$$

$$= 2\ x\ 46.52 - 58.15 = 34.89 \frac{kJ}{kmol.K} \quad (14)$$

$$F_{Ao} = u_o C_{Ao} = 0.472\ x\ 8.01\ x\ 60, \frac{m^3}{s}\frac{kmol}{m^3}\frac{s}{min}$$

$$= 226.84 \frac{kmol}{min} \quad (15)$$

$$F_{Io} = u_o C_{Io} = 0.472\ x\ 18.69\ x\ 60, \frac{m^3}{s}\frac{kmol}{m^3}\frac{s}{min}$$

$$= 529.30 \frac{kmol}{min} \quad (16)$$

$$\sum F_i Cp_i \Delta T_i$$

$$= (226.84\ x\ 58.15 + 529.3\ x\ 69.78)\ x\ (T - 300), \frac{kmol}{min}\frac{kJ}{kmol.K}.K.$$

$$= 50125.3(T - 300), \frac{kJ}{min} \quad (17)$$

$$X[\Delta H_R^T + \Delta Cp(T - T_R)]F_{Ao}$$

$$= 0.4[14,890 + 34.89(T - 300)]\ x\ 226.84$$

$$= 1351059.04 + 3165.779T - 949733.712$$

$$= 401325.328 + 3165.779T \quad (18)$$

Hence equating (17) and (18)

$$50125.3(T - 300) = 401325.328 + 3165.779T$$

or

$$50125.3T - 15037590 = 401325.328 + 3165.779T$$

from which T = 311.7 K

Putting T = 311.7 K in (12)

$$k_1 = 6.55e^{-1022/311.7} = 0.2468 \ min^{-1} \qquad (19)$$

From the van't Hoff equation

$$\frac{d\ln K}{dT} = \frac{\Delta H_R^T}{RT^2}$$

That is

$$\ln\frac{K_2}{K_1} = -\frac{\Delta H_R}{R}\left(\frac{1}{T_2} - \frac{1}{T_1}\right) \qquad (20)$$

At 232.6 K

$$\ln\frac{K_2}{K_1} = -\frac{14890}{8.314}\left(\frac{1}{311.7} - \frac{1}{300}\right) = 0.2241$$

$$\frac{K_2}{K_1} = 1.2512 \qquad (21)$$

Thus, since $K_{300} = K_1 = 160.19$, given

$$K_2 = 1.2512 \ x \ K_1 = 1.2512 \ x \ 160.19 = 200.43 \qquad (22)$$

From (10), (19) and (22)

$$\frac{k_1}{k_2} = K_2 \quad or \quad k_2 = \frac{0.2468 \ min^{-1}}{200.43} = 0.00123 min^{-1} \qquad (23)$$

Putting (19), (23) in (7),

$$-r_A = k_1 C_A - k_2 C_B^2$$

$$= 0.2468\frac{8.01(1 - X_A)}{(1 + 0.3X_A)} - 0.00123\left(\frac{16.02X_A}{(1 + 0.3X_A)}\right)^2$$

$$= \frac{1.977(1 - X_A)}{(1 + 0.3X_A)} - 0.317\left(\frac{X_A}{1 + 0.3X_A}\right)^2$$

$$= \frac{1.977(1 - X_A)(1 + 0.3X_A) - 0.317{X_A}^2}{(1 + 0.3X_A)^2}$$

$$b = \frac{1.977 - 1.384X_A - 0.91{X_A}^2}{(1 + 0.3X_A)^2} \qquad (24)$$

We can calculate $-r_A$ at various X_A as shown in the Table below.

X_A	$-r_A$
0.00	1.98
0.10	1.72
0.20	1.48
0.30	1.25
0.40	1.02
0.50	0.80
0.60	0.59

This is plotted in Fig. 5.6a as $-r_A$ versus X_A.

Fig. 5.6a: Plot of r_A versus X_A

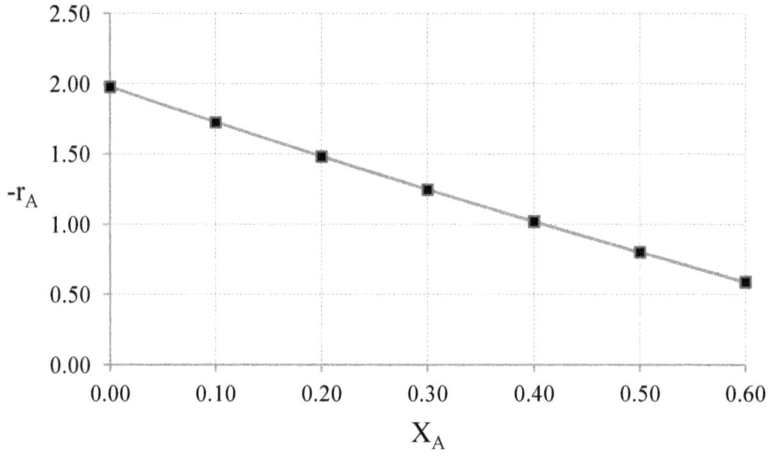

a. For a plug flow reactor at 40 % conversion

$$V = F_{Ao} \int_0^X \frac{dX_A}{-r_A}$$

$$= F_{Ao}\int_0^{0.4}\frac{(1+0.3X_A)^2 dX_A}{1.977-1.384X_A-0.91{X_A}^2} \quad (25)$$

The integral may be obtained numerically using Simpson's rule. Thus if we express equation (25) as

$$V = F_{Ao}\int_0^{0.4} Y dX_A \quad (25a)$$

where

$$Y = \frac{(1+0.3X_A)^2}{1.977-1.384X_A-0.91{X_A}^2} \quad (25b)$$

$$Integral = \frac{h}{3}(Y_o + 4Y_1 + Y_2) \quad (26)$$

where

$$h = \frac{X_{A2}-X_{Ao}}{2} \quad (27)$$

We can choose $X_{A2} = 0.20,\ X_{Ao} = 0$ for the first half of the numerical integration and $X_{A2} = 0.40,\ X_{Ao} = 0.20$ for the second half. A Table of Y versus X_A is shown below and plotted in Fig. 5.6b.

X_A	Y
0.00	0.51
0.10	0.58
0.20	0.68
0.30	0.80
0.40	0.98

Thus, by Simpson's rule, between $X_A = 0$ and $X_A = 0.4$,

$$V = 226.84\frac{kmol}{min}\left[\frac{0.1}{3}(0.51 + 4\,x\,0.58 + 0.68) + \frac{0.1}{3}(0.68 + 4\,x\,0.80 + 0.98)\right]\frac{m^3 min}{kmol}$$

$$= 63.29\, m^3 \quad Ans$$

Fig. 5.6b: Plot of Y versus X_A for Numerical Integration

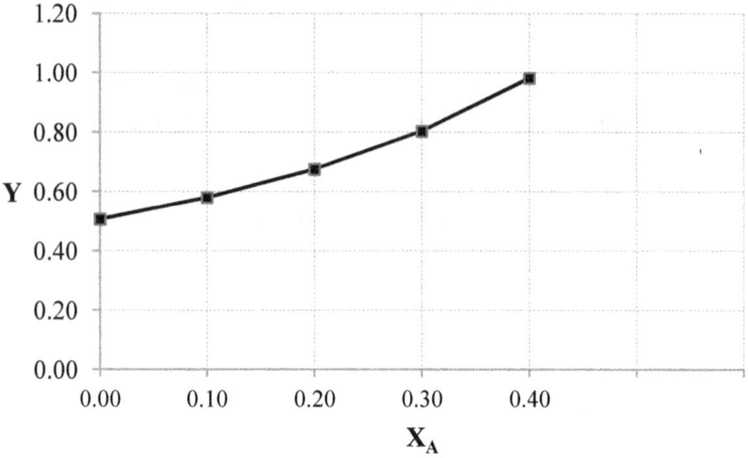

b. For a CSTR

$$V = \frac{F_{Ao} X_A}{-r_A} \tag{28}$$

At 40 % conversion, from equation (24)

$$-r_A = \frac{1.977 - 1.384X_A - 0.91{X_A}^2}{(1 + 0.3X_A)^2}$$

$$= \frac{1.977 - 1.384\ x\ 0.4 - 0.91\ x\ (0.4)^2}{(1 + 0.3\ x\ 0.4)^2} = 1.0187 \frac{kmol}{m^3 s} \tag{29}$$

From (15), (28) and (29)

$$V = \frac{F_{Ao} X_A}{-r_A} = \frac{226.84\ x\ 0.4}{1.0187} \frac{kmol}{s} \frac{m^3.s}{kmol} = 89.07\ m^3 \quad Ans$$

Example 5.7

The endothermic liquid phase elementary reaction

$$A + B \rightarrow 2C$$

proceeds, substantially, to completion in a single, steam jacketed, continuous, stirred tank reactor. From the following data, calculate the steady state reactor temperature

Reactor volume	0.5 m^3
Steam jacket area	0.929 m^2
Jacket steam	10.34 bar (185.5 C saturation temperature)
Overall heat transfer coefficient	852 W/m^2K
Agitator shaft horse power	25 HP
Heat of reaction, ΔH_R	+46,520 kJ/kmol (independent of temperature)

	A	B	C
Feed rate, kmol/h	4.54	4.54	0
Feed temperature, C	27	27	
Specific heat, kJ/kmol.K (independent of temperature)	213.54	184.23	198.88
Molecular weight	128	94	
Density, kg/m^3	1009.2	1076.5	1041.2

Answer

For the given reaction

$$A + B \rightarrow 2C$$

X_A may be taken to be equal to 1 as the reaction goes, substantially, to completion. There is also no volume change during reaction. The reactor may be represented, physically, as shown below.

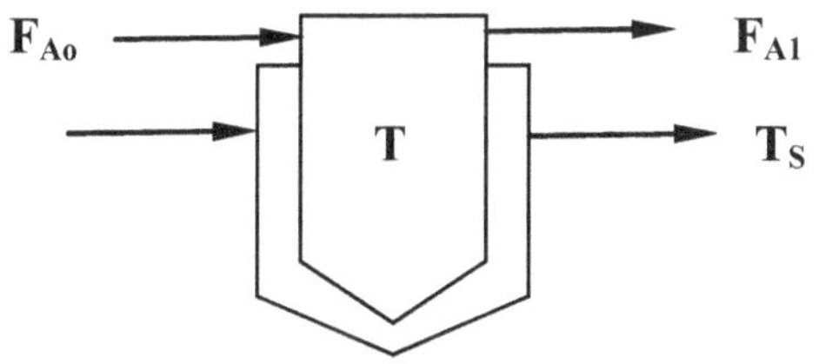

The energy balance across the reactor is

$$Q - F_{Ao}X_A[\Delta H_R + \Delta\overline{Cp}(T - T_R)] + W_S = F_{Ao}\sum_{i=1}^{n}\theta_i\overline{Cp_i}\Delta T_i \qquad (1)$$

where

$$\theta_i = \frac{F_{io}}{F_{Ao}} \text{ and } \Delta T_i = T - T_{io}$$

From the given numerical values

$$\Delta\overline{Cp} = 2\,x\,198.88 - 213.54 - 184.23 = -0.01\frac{kJ}{kmol.K}$$

$$W_S = 25\,x\,0.746 = 18.65\ \ kJ/s$$

$$Q = UA(T_S - T_R) = 0.852\,x\,0.929\,(458.5 - T_R), \frac{kW}{m^2K}.m^2.K$$

$$= 0.792(458.5 - T_R)\frac{kJ}{s}$$

$$F_{Ao} = \frac{4.54}{3600} = 1.26\,x\,10^{-3}\ kmol/s$$

$$\theta_A = \frac{F_{Ao}}{F_{Ao}} = \frac{1.26\,x\,10^{-3}}{1.26\,x\,10^{-3}} = 1$$

$$\theta_B = \frac{F_{Bo}}{F_{Ao}} = \frac{1.26\,x\,10^{-3}}{1.26\,x\,10^{-3}} = 1$$

$$\theta_C = \frac{F_{Co}}{F_{Ao}} = \frac{0}{1.26\,x\,10^{-3}} = 0$$

Substituting these values in equation (1)

$$0.792(458.5 - T_R) - 1.26\,x\,10^{-3}\,x\,1[46{,}520 - 0.01(300 - T_R)] + 18.65$$
$$= 1.26\,x\,10^{-3}(1\,x\,213.54 + 1\,x\,184.23 + 0\,x\,198.88)(T_R - 300)$$

That is

$$363.132 - 0.792T_R - 58.615 + 0.00378 - 1.26\,x\,10^{-5}$$
$$= 0.501T_R - 150.357$$

$$458.878 = 1.293T_R$$

$$T_R = 354.894\ K\ (81.9\ C)\ Ans$$

Example 5.8

The second order, irreversible, gas phase reaction

$$A + B \rightarrow C$$

is carried out, isothermally, at 815 K and 5 atm, in a 101.6 mm diameter, tubular, plug flow, reactor. The feed rate is 9.08 kmol/h, consisting of stoichiometric amounts of reactants A and B and 20 mole % inerts, I.

a. At what distance along the reactor will the conversion be 30 %?
b. What will the wall heat flux be at this position if the endothermic heat of reaction is 53, 498 kJ/kmol at 815 K?

The second order reaction rate constant is k = 1.8714 x 10^4 m^3/h.kmol and the specific heats are

$$Cp_A = 1.047\frac{kJ}{kmol.K}$$
$$Cp_B = 1.256\frac{kJ}{kmol.K}$$
$$Cp_C = 0.837\frac{kJ}{kmol.K}$$
$$Cp_I = 1.675\frac{kJ}{kmol.K}$$

(Fogler, 1974)

Answer

<u>Part a.</u>

For a plug flow reactor

$$V = \int_0^{X_A} \frac{F_{Ao} dX_A}{-r_A} \tag{1}$$

For the given second order reaction

$$-r_A = kC_A C_B = kC_A^2 \tag{2}$$

In order to obtain the correct value of C_A, we need to construct, among other calculations, the stoichiometric table which, for this reaction, is

Specie	Moles initially present	Moles reacted	Moles remaining
A	F_{Ao}	$-F_{Ao}X_A$	$F_{Ao}(1 - X_A)$
B	F_{Bo}	$-F_{Ao}X_A$	$F_{Ao}(1 - X_A)$
C	F_{Co}	$F_{Ao}X_A$	$F_{Ao}X_A$
I	F_{Io}	0	F_{Io}
Totals	$F_{Ao} + F_{Bo} + F_{Co} + F_{Io}$	$-F_{Ao}X_A$	$F_{Io} + F_{Ao}(2 - X_A)$

$$\delta = 1 - 1 - 1 = -1 \tag{3}$$

Initial total feed rate = 9.08 $kmol/h$

Initial feed rate of A and B = 0.8 x 9.08 = 7.264 $kmol/h$

Initial feed rate of A, $F_{Ao} = \frac{7.264}{2} = 3.632\ kmol/h$

Initial mole fraction of A, $y_{Ao} = \frac{3.632}{9.08} = 0.400$ (4)

Thus, from (3) and (4), $\epsilon = y_{Ao}\delta = -0.4$

The volumetric rate of reaction mixture per unit time is

$$u = u_o(1 + \epsilon X_A) = u_o(1 - 0.4X_A), \frac{m^3}{h}$$

Thus, the concentration of A is

$$C_A = \frac{F_{Ao}(1 - X_A)}{u_o(1 - 0.4X_A)}, \frac{kmol}{h} \cdot \frac{h}{m^3} = \frac{C_{Ao}(1 - X_A)}{(1 - 0.4X_A)}, \frac{kmol}{m^3} \tag{5}$$

where u_o is the volumetric flow rate, m^3/h, entering the reactor.

For an ideal gas, since $R = 8.206 \; x \; 10^{-2} \; \frac{m^3 atm}{kmol.K}$

$$Pu_o = F_{Ao}RT$$

$$= 3.632 \; x \; 8.206 \; x \; 10^{-2} x \; 815, \frac{kmol}{h} \cdot \frac{m^3 atm}{kmol.K} . K$$

$$= 242.904. \frac{m^3 atm}{h}$$

from which, at 5 atm.,

$$u_o = \frac{242.904}{5}, \frac{m^3 atm}{h.atm} = 45.581 \; \frac{m^3}{h}$$

and

$$C_{Ao} = \frac{F_{Ao}}{u_o} = \frac{3.632}{45.581}, \frac{kmol}{h} \cdot \frac{h}{m^3} = 0.0797 \; \frac{kmol}{m^3}$$

Substituting for C_{Ao} in (5)

$$C_A = \frac{0.0797(1 - X_A)}{(1 - 0.4X_A)} \; \frac{kmol}{m^3} \tag{6}$$

Substituting (6) in (2)

$$-r_A = kC_A^2 = 1.8714 \; x \; 10^4 \left(\frac{0.0797(1 - X_A)}{(1 - 0.4X_A)}\right)^2, \frac{m^3}{h.kmol} \frac{(kmol)^2}{m^6}$$

$$= 118.873 \left(\frac{(1 - X_A)}{(1 - 0.4X_A)}\right)^2, \; \frac{kmol}{h \, m^3} \tag{7}$$

Substituting (7) in (1)

$$V = \frac{3.632}{118.873}\int_0^{0.3} \frac{(1-0.4X_A)^2 dX_A}{(1-X_A)^2},, \frac{kmol}{h}, \frac{h\,m^3}{kmol}$$

$$= 0.0306\int_0^{0.3} \frac{(1-0.4X_A)^2 dX_A}{(1-X_A)^2} m^3 \qquad (8)$$

But

$$V = \frac{\pi d^2 L}{4} \quad \text{from which} \quad L = \frac{4V}{\pi d^2}$$

Thus

$$L = \frac{4\,x\,0.0306}{\pi(0.1016)^2}\int_0^{0.3} \frac{(1-0.4X_A)^2 dX_A}{(1-X_A)^2} \frac{m^3}{m^2}$$

$$= 3.774\int_0^{0.3} \left(\frac{1-0.4X_A}{1-X_A}\right)^2 dX_A, m \qquad (9)$$

Equation (9) can be integrated numerically using Simpson's rule. Table 5.8 below shows the value of the integrand at various values of X_A.

Table 5.8: Values of the Integrand vs X_A

X_A	$\left(\frac{1-0.4X_A}{1-X_A}\right)^2$
0	1
0.075	1.099664
0.15	1.222976
0.225	1.37873
0.3	1.580408

These are plotted in Fig. 5.8

By Simpson's rule, area under the curve between the given limits of X_A

$$= \frac{0.075}{3}[(1 + 4\,x\,1.0997 + 1.2230) + (1.2230 + 4\,x\,1.3787 + 1.5804)]$$

$$= \frac{0.075}{3} x14.94 = 0.3735 \quad (10)$$

Substituting (10) in (9)

$$L = 3.774\ x\ 0.3735 = 1.41\ m\ Ans$$

Fig. 5.8: Plot of the Integrand versus X_A

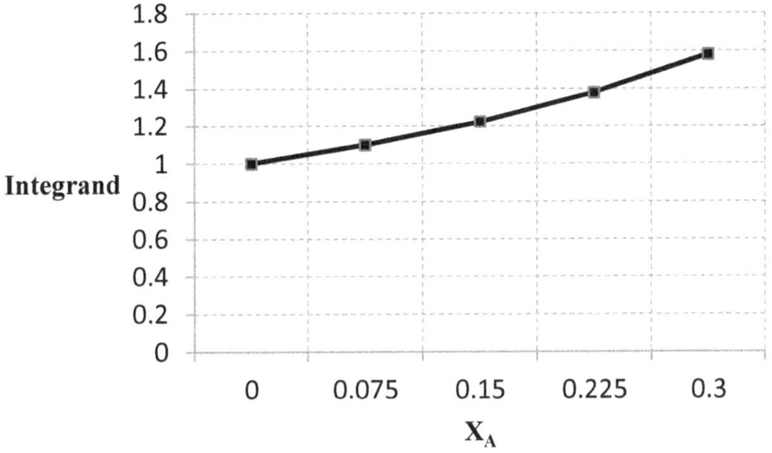

<u>Part b</u>

Since the reaction is isothermal throughout, the general energy equation

$$Q - F_{Ao}X_A[\Delta H_R + \Delta\overline{Cp}(T - T_R)] + W_S = F_{Ao}\sum_{i=1}^{n}\theta_i\overline{Cp_i}\Delta T_i$$

reduces to

$$Q - F_{Ao}X_A\Delta H_R \quad (11)$$

because $(T - T_R) = 0, W_S = 0\ and\ \Delta T_i = 0.$

Using the given values in equation (11)

$$Q = F_{Ao}X_A\Delta H_R = 3.632\ x\ 0.3\ x53.498\ , \quad \frac{kmol}{h}\cdot\frac{kJ}{kmol}$$

$$= 3.632\ x\ 0.3\ x53{,}498 = 58{,}291\ \frac{kJ}{h}$$

The heat flux

$$= \frac{Q}{Wall\ Area} = \frac{Q}{\pi DL} = \frac{58{,}291}{\pi\ x\ 0.1016\ x\ 1.41} = 129{,}520.59\ \frac{kJ}{h\,.m^2}$$

It is more usual to report this as

$$\text{Heat flux} = \frac{129{,}520.59\ \frac{kJ}{h\,.m^2}}{3600\frac{s}{h}} = 35.978\ \frac{kW}{m^2} \quad Ans.$$

References for Chapter Five

1. B. N. Nnolim: Unpublished Lecture Notes in Chemical Reaction Engineering; IMT, Enugu, Nigeria; 1989
2. Denbigh K. G. & Turner J. C. R.; *Chemical Reactor Theory*; 2nd edition, London; Cambridge University Press, 1971
3. Fogler, H Scott, *The Elements of Chemical Kinetics and Reactor Calculations*; Prentice Hall Inc; N. J., USA, 1974
4. Levenspiel O; *Chemical Reaction Engineering*; Wiley International Edition; New York, USA, 1972
5. Walas S. M.; *Reaction Kinetics for Chemical Engineers*; McGraw-Hill Book Company, New York, USA, 1959

www.ingramcontent.com/pod-product-compliance
Lightning Source LLC
Chambersburg PA
CBHW031931190326
41519CB00007B/485
9781906914196